Make: Planes, Gliders, and Paper Rockets

*Simple Flying Things Anyone
Can Make—Kites and Copters, Too!*

Rick Schertle and James Floyd Kelly

MAKER MEDIA™
SAN FRANCISCO, CA

Make: Planes, Gliders, and Paper Rockets

by Rick Schertle and James Floyd Kelly

Printed in the United States of America.

Published by Maker Media, Inc., 1160 Battery Street East, Suite 125, San Francisco, CA 94111.

Maker Media books may be purchased for educational, business, or sales promotional use. Online editions are also available for most titles (*http://safaribooksonline.com*). For more information, contact our corporate/institutional sales department: 800-998-9938 or *corporate@oreilly.com*.

Editor: Patrick Di Justo
Production Editor: Matthew Hacker
Copyeditor: Jasmine Kwityn
Proofreader: Phil Dangler

Indexer: WordCo Indexing Services, Inc.
Interior Designer: David Futato
Cover Designer: Brian Jepson
Illustrator: Rebecca Demarest

November 2015:   First Edition

Revision History for the First Edition

2015-10-16   First Release

See *http://oreilly.com/catalog/errata.csp?isbn=9781457187698* for release details.

978-1-4571-8769-8

[LSI]

Table of Contents

Preface

Welcome to the first in our series of maker books: *Planes, Gliders, and Paper Rockets: Simple Flying Things Anyone Can Make—Kites and Copters, Too!* We are so excited to be creating this series of books for Maker Media, and it is our hope that readers (young and old) will find the subject matter and projects in each book inspirational and motivating.

Making things is fun! Students don't need to tell you that—you (as a teacher or parent) can just see the smiles on their faces as they cut, glue, tighten, and tape. Not only that, making things allows students to continue to learn new skills and develop a deeper understanding of the world around them. But as parents and teachers, sometimes thinking about how things in our great big world work and how best to teach these concepts can be a little overwhelming. It often feels like there is just so much to learn and questions related to the topic we wish to explore begin to appear:

- Where do I start and how do I start with a particular topic?
- What is important in grasping the basics of a topic and what is not important?
- What activities or tasks can I have a student perform that will make comprehending a topic easier and faster?
- What tools and supplies will I need on hand and in what quantities to ensure all students are engaged?

Hands-on activities have long been known to be successful when teaching new topics, and teachers and parents know that students are more likely to pay attention and retain new knowledge when their hands are as engaged as their eyes and ears. And for that reason, this book series was created.

This series of books will provide parents and teachers with a fun mix of projects, discussion materials, instructions, and subjects for deeper investigation. Our goal is to help you spend more time learning and experimenting and less time planning and preparing the lessons you wish to teach. You'll find lists of required tools and supplies in one place for all of the book's projects. We'll provide detailed instructions for projects (including photos), downloads of PDF templates for specific steps, and videos of several projects in operation. And finally, we will provide discussion material, questions, and suggestions for challenges you can present to your students for taking projects to the next level.

From the start, we wanted to create a book series that we wish our teachers and parents would have had access to when we were students. We hope we've been successful in that endeavor and that you enjoy the series, and we welcome your questions, comments, and suggestions. If you have ideas for additional book topics or suggestions for improving the series, please let us know.

Overview

Here you will find a short summary of all the projects found in the book. More detailed coverage that includes step-by-step instructions and a suggested structure for opening a discussion and moving it in the direction of a particular hands-on project will all be provided in the individual chapters of the book, which are each dedicated to a specific theme or subject. Keep in mind that you don't have to perform these projects in the order we provide them. Pick the ones you need, skip the ones your children or students aren't ready for or you're not yet prepared to tackle, and have fun!

Chapter 1, Helicopters

There's just something really fun and cool about helicopters—maybe it's the way they can just hover there, staying in one place unlike airplanes and other flying contraptions. Studying helicopters opens up a wide range of discussion topics: where are their wings, for example? Examining helicopters also means looking into airflow, gravity, torque, power, and force, just to mention a few. In this chapter, you'll build a simple paper catapult helicopter that flutters down after being shot into the air, and then you'll make a pull-cord stick helicopter that achieves lift by pulling a cord to get its wings spinning fast.

Chapter 2, Rockets

What book on things that fly would be complete without a chapter on rockets? Definitely not this one! Everyone loves rockets, and the projects in this chapter deliver a number of hands-on activities that are sure to ignite an interest in space, astronauts, satellites, and obviously, the Universe. We'll also discuss topics like ballistics, pressure, and gravity. You'll build a low-pressure launcher and rocket, and when you're ready to go higher, a high-pressure launcher and rocket. These projects work well with individuals, but are perfect for groups. What's more, the rockets and launchers you build can be used over and over again.

Chapter 3, Airplanes and Gliders

Sure you can buy a rubber band and propeller airplane in the stores, but that's not as fun as making your own! You'll find complete instructions for a super easy to build airplane that will amaze you with its flight capabilities, as you also learn about balance and airflow. And how much fun will it be to combine two of this book's subjects into one project: a rocket glider! The rocket goes up, wings pop out, and the whole thing glides back down. This one is guaranteed to impress.

Chapter 4, Rockets Revisited

We're not done with rockets, and hopefully you're not either! There are so many more rocket projects for your consideration—a rocket that bounces off the ground when it returns from its launch and even a rocket with its own parachute. And while the paper and tape rockets are popular (and you'll discover steps to build a paper and tape rocket building stand in this chapter as well), you're going to love the foam rocket option that is more durable with different flight characteristics.

Chapter 5, Kites!

When it comes to flying things, it really doesn't get much simpler than a kite—a breeze or stiff wind does all the work. All you've got to do is make sure to hold on to the string! Here are two simple kite projects that are sure to delight, whether at the beach or local park or with an individual or a group.

Tools and Supplies

This section lists the tools and supplies needed for all of the projects in this book. Rather than provide these lists for each chapter, we feel you'll spend less time hunting for these items if you're given the complete list from the start.

We still recommend reading over each project beforehand to better understand the construction process for each project; this will also help ensure you collect the right tools ahead of time along with any additional tools or supplies you feel you may need for a project.

Tools

- Scissors
- Coping saw
- Glue gun and sticks
- Drill and drill bits of various sizes
- Safety glasses
- Hacksaw or PVC cutter
- Hair dryer
- Bench vise or clamp
- Single hole punch
- Ruler
- Meter stick
- Pens/pencils
- Stapler
- Hobby knife
- Pliers (needle nose and slip joint)
- Adjustable wrench
- Bicycle pump

Supplies for Paper Catapult Helicopter

- Rubber bands, #16 recommended (1 per student)
- Large paper clip (1 per student)
- Printed paper template, download from *http://www.airrocketworks.com/make-planes-gliders-rockets*
- Clear tape
- 1/4" diameter wood dowel, 6" long (1 per student)

- LED, any color (1 per student)
- Coin battery (1 per student)

Supplies for Pull-Cord Stick Helicopter

- Jumbo craft sticks (6" × 3/4")
- 1/4" diameter wood dowel, 6" long
- Fine/strong nylon kite string (3' per student)
- Permanent marker
- Pens for decorating
- Match to burn end of string
- Small stick scrap for pull string handle
- 1/2" PVC pipe, 6" piece

Supplies for Low-Pressure Rocket Launcher and Rockets

- 10' of 1/2" Schedule 40 PVC cut to the following lengths for one launcher: 3' (1x), 2' (1x), 1' (2x), 3" (1x)
- Duct tape
- 1/2" PVC slip tees (2 per launcher)
- Scraps of paper
- PVC conduit 90 degree elbow (1 per launcher)
- Brightly colored 8.5" × 11" paper (or plain white if you want to decorate yourself)
- PVC pipe, 1/2" × 12" (to be used as the form to build rockets)
- Filter mask (1 per student)
- Clear or masking tape
- Wad of clear plastic wrap
- 2-liter soda bottle (one per launcher)

Supplies for High-Pressure Rocket Launcher and Rockets

- Compressed Air Rocket Launcher v2.1 kit (purchased from Maker Shed (*http://bit.ly/car_launcher_v2_1*) or AirRocketWorks.com.
- Masking tape (3/4" wide or similar)
- PVC pipe - 1/2" x 12" (to be used as the form to build rockets)

- Template for high-pressure rocket (*http://bit.ly/mpgpr_templates*)

Supplies for Rubber Band Airplane

- 9" foam plates (3)
- Small paper clip
- Clear tape
- Two fat, 8" soda straws (the kind for smoothies or milkshakes)
- Rubber bands, 28" buy 1/8" wide × 50' long from Kelvin.com for about $7.49
- 6" nose hook propeller, about $0.49 at Kelvin.com
- Printed wing templates (*http://bit.ly/mpgpr_templates*)
- Electric rubber band winder (handy, but not necessary), about $5 from *http://www.hobbyking.com*

Supplies for Air Rocket Glider

- The Air Rocket Glider can be built from scratch, but it's much easier to build from a kit available at: *http://www.airrocketworks.com*.
- Super glue

Supplies for Foam Air Rocket

- Packing tape, clear roll (1)
- Duct tape (1 roll), now available in a variety of fun colors and designs
- Zip tie, 8' (1)
- Foam sheet, 2-mm thick, (1), available at craft stores or online
- Foam pipe insulation, 1/2' inside diameter (1); you can build 8 rockets with a 6' piece (instructions here are for one rocket).

Supplies for Rocket Building Stand

- 1/2" PVC pipe, 15" long
- 2" × 6" lumber
- 1 5/8" wood screw

Supplies for Bounce Rocket

- 27-mm super bouncy rubber ball
- Nylon tube (11/16" inside diameter, 3/4" outside diameter), available from *http://bit.ly/nylon_tube*
- 3D-printed fins (download STL files at *http://bit.ly/mpgpr_templates*)
- Three Fin Guide (*http://bit.ly/mpgpr_templates*)

Supplies for Air Rocket with Parachute (ARP)

- Plastic Easter egg
- Jumbo craft stick (6" × 3/4")
- 8" long piece of rubber band
- Plastic milk jug
- Scrap of card stock
- Small, flat washer (.625" OD or similar)
- Plastic bag
- Lightweight string
- Clothes hanger or similar gauge wire
- Masking tape (3/4" wide or similar)
- PVC pipe, 1/2" × 12" (to be used as the form to build rockets)
- Template for high-pressure rocket (*http://bit.ly/mpgpr_templates*)

Supplies for Quick-Build Kite for $0.25

- Brightly colored 8.5" × 11" paper
- Masking tape
- 7.5" long wood or plastic coffee stirrer
- Small piece of cardboard (for wrapping kite string)
- Lightweight nylon kite string (the length needed will depend on how high you want your kite to fly)
- Clear tape
- Flagging tape (nonadhesive plastic ribbon), available in the surveying section of hardware stores

Supplies for Simple Sled Kite

- Lightweight nylon kite string
- 1/4" diameter wood dowels, 250' (2)
- Tyvek® 3' × 4' (can purchase by the foot on eBay)
- Tyvek® tape
- Flagging tape (nonadhesive plastic ribbon), optional

Conventions Used in This Book

The following typographical conventions are used in this book:

Italic

> Indicates new terms, URLs, email addresses, filenames, and file extensions.

`Constant width`

> Used for program listings, as well as within paragraphs to refer to program elements such as variable or function names, databases, data types, environment variables, statements, and keywords.

`Constant width bold`

> Shows commands or other text that should be typed literally by the user.

This element signifies a general note, tip, or suggestion.

This element indicates a warning or caution.

Safari® Books Online

Safari Books Online is an on-demand digital library that delivers expert content in both book and video form from the world's leading authors in technology and business.

Technology professionals, software developers, web designers, and business and creative professionals use Safari Books Online as their primary resource for research, problem solving, learning, and certification training.

Safari Books Online offers a range of plans and pricing for enterprise, government, education, and individuals.

Members have access to thousands of books, training videos, and prepublication manuscripts in one fully searchable database from publishers like Maker Media, O'Reilly Media, Prentice Hall Professional, Addison-Wesley Professional, Microsoft Press, Sams, Que, Peachpit Press, Focal Press, Cisco Press, John Wiley & Sons, Syngress, Morgan Kaufmann, IBM Redbooks, Packt, Adobe Press, FT Press, Apress, Manning, New Riders, McGraw-Hill, Jones & Bartlett, Course Technology, and hundreds more. For more information about Safari Books Online, please visit us online.

How to Contact Us

Please address comments and questions concerning this book to the publisher:

> Make:
> 1160 Battery Street East, Suite 125
> San Francisco, CA 94111
> 877-306-6253 (in the United States or Canada)
> 707-639-1355 (international or local)

Make: unites, inspires, informs, and entertains a growing community of resourceful people who undertake amazing projects in their backyards, basements, and garages. Make: celebrates your right to tweak, hack, and bend any technology to your will. The Make: audience continues to be a growing culture and community that believes in bettering ourselves, our environment, our educational system—our entire world. This is much more than an audience; it's a worldwide movement that Make: is leading—we call it the Maker Movement.

For more information about Make:, visit us online:

Make:magazine: *http://makezine.com/ magazine*
Maker Faire: *http://makerfaire.com*
Makezine.com: *http://makezine.com*
Maker Shed: *http://makershed.com*

We have a web page for this book, where we list errata, examples, and any additional information. You can access this page at *http://bit.ly/ planes_gliders_paper_rockets*.

To comment or ask technical questions about this book, send email to *bookquestions@oreilly.com*.

Acknowledgments

This being my first book, I'm grateful to my expert coauthor Jim, who helped walk me through. I'm thankful to my dad, Bill (a maker himself), who allowed me to work beside him as a kid and my mom, Anita, who was OK with us making a mess. We dabbled in many things and became experts at none, but had a lot of fun along the way. Thanks to Mark Frauenfelder who gave me a chance with my first project for *Make: Magazine* back in 2008. Thanks to Dale, Sherry, and Louise at Maker Media for bringing a name and face to what many of us have done all our lives. My kids have now grown up with

Maker Faire! Thanks to Keith Violette, my business partner at Air Rocket Works, for being a constant source of design brilliance in dreaming up new flying things. Thanks to Slater Harrison, Nick Drogota at Howtoons, Jonathan at Big Wind Kites, Tim Parish at MyBestKite.com, Bill Kuhl, and Mike Westerfield. Thanks to John Olsen and my colleagues and students at Price Middle School. Most of all, thanks to my wife, Angie, and to my kids, Kelly and Micah, who are a continual source of encouragement and inspiration.

—Rick Schertle

I'd like to give special thanks to my coauthor, Rick, for being such a great partner to work with on this new book. I met Rick years ago at the Bay Area Maker Faire and we've maintained communication and have always been looking for a chance to work together. It finally happened.

My own experiences as a maker are a direct result of two parents who were both makers. So special thanks to Donna and Darrell Kelly for raising me in a house where there was always something being built or fixed or taken apart.

A big thank you goes to Brian Jepson, who took Rick's and my idea and helped make it happen. I always enjoy visiting with Brian wherever we run into one another, and if you like this book series, be sure to let him know via Maker Media.

—James Floyd Kelly

Helicopters

An airplane by its nature wants to fly....A helicopter does not want to fly. There is no such thing as a gliding helicopter.

—Harry Reasoner, US broadcast journalist

There is something amazing when you really look at a helicopter hovering in the sky. Airplanes must keep moving forward so that the airflow over the wings provides lift and keeps it in the sky. But helicopters behave differently, don't they?

What can we observe about a hovering helicopter?

- Its largest blade is parallel to the ground, as opposed to an airplane's propeller being perpendicular.
- The small blade on the rear of the helicopter is perpendicular to the ground, but it points away from the side of the helicopter instead of to the front like an airplane.

Now consider the following questions:

- Why do the blades on a helicopter have to be at a slight angle? Explore how an airplane's wings get lift and then try to determine what the "wing" on a helicopter is and how it creates lift.

- Do helicopters take more or less energy to stay in the sky than an airplane?
- What are things helicopters can do that airplanes cannot?

Paper Catapult Helicopter

This simple little paper and tape helicopter project has been popular for a long time and done with many different styles. The traditional copters can easily be thrown a few feet in the air for a few feet of flight. For a slightly longer flight, it can be dropped from a second story balcony. But there's actually a much easier way to get the helicopter up to a good height—just create a rubber band catapult! Using a notched dowel or your finger, you can fire this little copter 30' in the air for long flights down! For our helicopter, we've also chosen to add an LED (and battery) that adds some nice flash to the project. Even better, the helicopter can be flown at night so that an interesting lighting effect is produced as it comes down.

But don't stop there! Each chapter in this book (and future books) will have multiple projects. For the second helicopter project, we'll be building a pull-string stick heli, a fun little project that uses PVC, a material you might not have worked with before. We'll also examine how a person's muscles can be used to transfer energy to the helicopter.

Let's take a look first at what's needed to build the basic paper helicopter (seen in Figure 1-1) and rubber band launcher.

Build the Paper Catapult Helicopter

The following steps each include a photo that can be used by the instructor or student to verify a task before moving on to the next step.

Step 1

Gather the supplies shown in Figure 1-2. You can find a complete list of the materials required for this project in "Tools and Supplies".

Figure 1-1 *The paper catapult helicopter.*

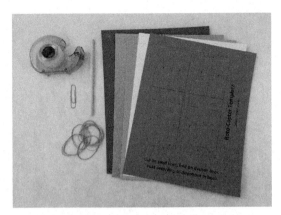

Figure 1-2 *Basic supplies for paper catapult helicopter.*

Step 2

Cut out the paper pieces, making sure to cut only on the solid lines, as shown in Figure 1-3.

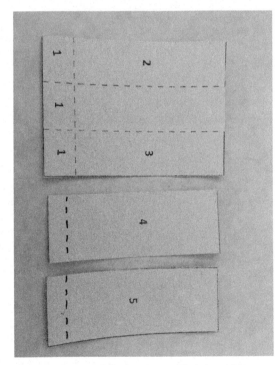

Figure 1-3 *Three paper cutouts will make up the helicopter body.*

Step 3

On the larger piece of paper labeled with the three 1s and 2 and 3, fold on the dotted line just above the 1s, as shown in Figure 1-4.

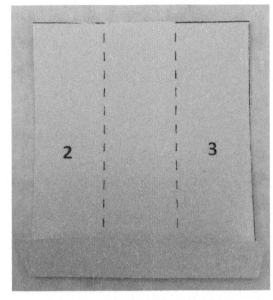

Figure 1-4 *Fold on the dotted line above the 1s.*

Step 4

Insert a paper clip in the middle row of the paper (between rows 2 and 3) so that it clips on the flap created in step 3. Use a piece of clear tape to hold the paper clip in place, as shown in Figure 1-5.

Step 5

Fold row 2 to the right (along the dotted line), as shown in Figure 1-6.

Step 6

Fold row 3 to the left (along the dotted line), turn sideways, and tape both ends closed to create the main body, as shown in Figure 1-7.

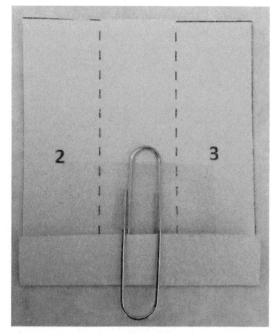

Figure 1-5   *Tape a paper clip over the folded section.*

Step 7

Take the piece of cut paper with the numeral 4 on it and lightly bend it on the dotted line. Do not fold it completely. Tape this piece onto the main body you made in step 6 at a slight angle to the left, as shown in Figure 1-8.

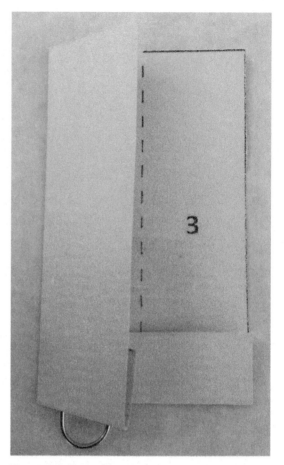

Figure 1-6   *Fold on the dotted line to the right of the numeral 2.*

Figure 1-7   *Fold again on the dotted line to the right of the numeral 3.*

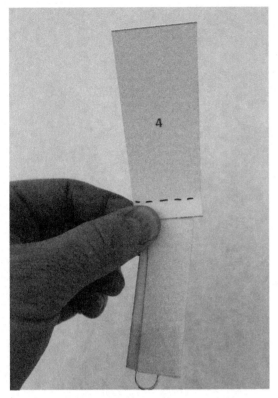

Figure 1-8 *Tape the main body to cutout 4.*

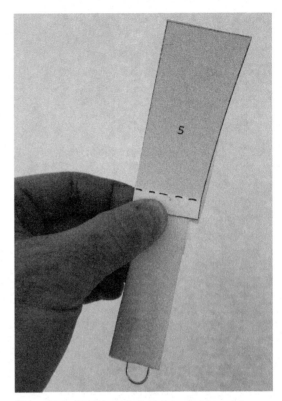

Figure 1-9 *Flip the main body over and tape it to cutout 5.*

Step 8

Flip the main body over. Take the piece of cut paper with the numeral 5 on it and lightly bend it on the dotted line. Do not fold completely. Tape this piece to the main body at a slight angle to the left, as shown in Figure 1-9.

Step 9

Bend blade 4 and blade 5 further down and place a piece of clear tape across the two blades so that they are perpendicular to the main body, as shown in Figure 1-10.

Step 10

Thread the rubber band through the exposed part of the paper clip, as shown in Figure 1-11.

Figure 1-10 *Fold down two blades (4 and 5) in opposite directions, and tape across the top.*

Figure 1-11 *Insert rubber band through paper clip.*

Step 11

Loop one end of the rubber band through the other end and pull tight to secure the rubber band, as shown in Figure 1-12.

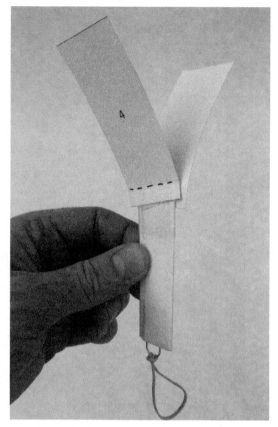

Figure 1-13 *The completed paper catapult helicopter.*

Build the Helicopter Launcher

To make the helicopter fly, it needs a launcher.

Step 1

Insert a 6″ length of dowel into a vise and cut a notch in one end using a coping saw, as shown in Figure 1-14.

Step 2

Insert the end of the rubber band into the notch. Hold the dowel with one hand while holding the helicopter (with wings folded down against the main body) in the other. Hold the dowel up and pull down on the helicopter (toward your body), as shown in Figure 1-15.

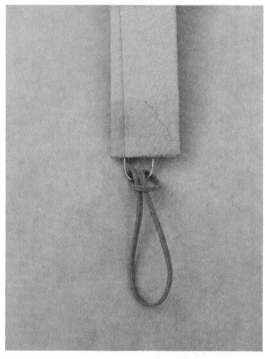

Figure 1-12 *Pull one end of rubber band through other end.*

Step 12

Figure 1-13 shows the completed paper catapult helicopter.

Figure 1-14 *Hold dowel in a vise grip before cutting notch.*

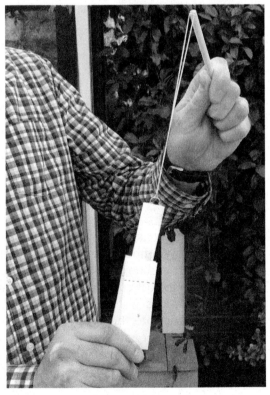

Figure 1-15 *Point dowel toward the sky, insert rubber band, and pull down.*

Step 3

Release the helicopter and it should fly high into the sky before the blades open up. The blades will spin as the paper helicopter returns safely to the ground, as shown in Figure 1-16. If the blades don't fold out like they should, adjust the angle of blades 4 and 5 until you get great flights.

Figure 1-16 *The blades open, and the helicopter spins as it returns to the ground.*

Add an LED for Night Launch

Once you're happy with your paper helicopter, there's an easy upgrade that can be added to allow for night launches.

Step 1

Gather the LED and coin battery (and some clear tape), as shown in Figure 1-17.

Step 2

Tape the longer "leg" of an LED to the positive side of the battery (look for the positive + on one side of the coin battery; the flip side is negative -) and wrap the tape around to complete the circuit, as shown in Figure 1-18.

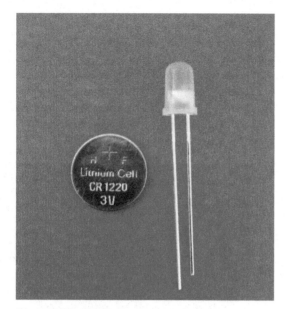

Figure 1-17 *An LED and a coin battery*

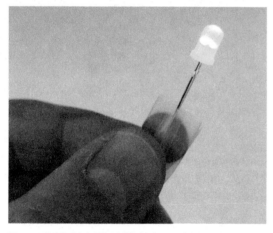

Figure 1-18 *Light the LED by taping legs to the coin battery.*

✏️ *Examine the LED*

Look closely at the LED's two "legs": these are two wires that have two different lengths. For electronics components, a wire "leg" like the one on the LED is typically called a lead (rhymes with "feed"). Most components have at least two leads, although some have more.

Step 3

Tape the LED and battery to your helicopter, as shown in Figure 1-19.

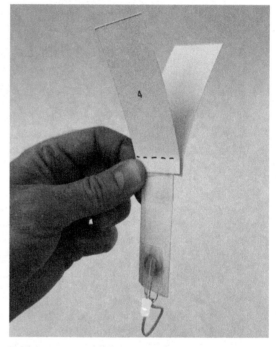

Figure 1-19 *LED and coin battery taped to the main body.*

Pull-String Stick Helicopter

This pull-string stick copter is also a classic made from easy-to-find materials. The materials for this project are inexpensive; if you're working with a group, just cut enough for everyone. After building, kids will have a blast sending these copters up into the sky at the local park or school yard. You can see an example of a completed pull-string copter in Figure 1-20.

Build the Pull-String Stick Helicopter

Step 1

Gather the supplies shown in Figure 1-21.

Figure 1-20 *The finished pull-string stick helicopter lifting out of the launch tube!*

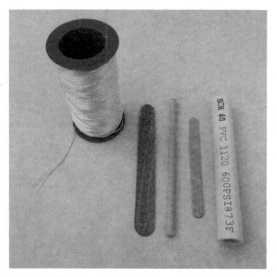

Figure 1-21 *The supplies for the pull-string stick helicopter.*

Step 2

Mark a jumbo 6" craft stick in the middle (at 3") with a permanent maker, as shown in Figure 1-22.

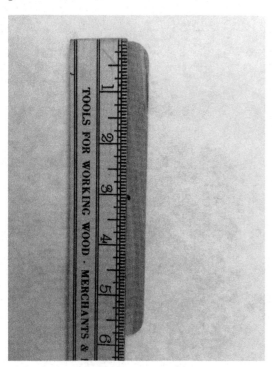

Figure 1-22 *Make a mark in the middle of a craft stick.*

Step 3

Soak a craft stick in water for 15 minutes, as shown in Figure 1-23.

Figure 1-23 *Soak the craft stick in water.*

Step 4

Place the soaked stick on a scrap of wood and carefully drill a 1/4" hole on the mark, as shown in Figure 1-24.

Figure 1-24 *Drill a hole at the mark (middle of craft stick).*

Step 5

Now you're going to shape your helicopter blade. You'll need two people for this one. One person should twist the wet stick with the left side of the stick twisted up and the right twisted down. While the first person continues to hold the stick in the twisted position, the second person should use a hair dryer for a few minutes to dry the stick until it's set, as seen in Figure 1-25.

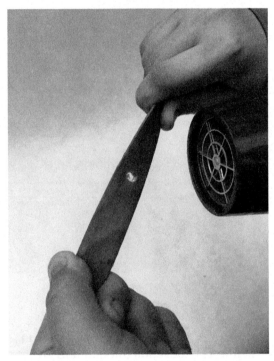

Figure 1-25 *Use a hair dryer to dry the stick while twisting the ends.*

Step 6

Cut a 1/4" diameter dowel to 6" using a coping or hack saw. Hold dowel in vice while cutting, as shown in Figure 1-26.

Figure 1-26 *Use a saw to cut a dowel with a length of 6".*

Step 7

From one end of the dowel (we'll call it the top end), measure down 1" and carefully drill a 1/4" hole, as seen in Figure 1-27.

Figure 1-27 *Drill a hole in the dowel.*

Step 8

Push the dowel through the hole in the craft stick and add hot glue on the top and bottom. You can see two different views of the helicopter in Figures 1-28 and 1-29.

Figure 1-28 *Top of craft stick with dowel secured with hot glue.*

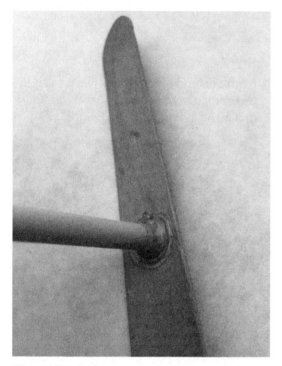

Figure 1-29 *Underside of craft stick with dowel secured by hot glue.*

Your stick helicopter is done; next, we'll build the pull-cord mechanism.

Step 9

With the PVC pipe held securely in a vice, and wearing a mask to protect against the fumes produced by cutting or drilling PVC,

use your hacksaw to cut a 6" section (or use PVC cutters), as marked in Figure 1-30.

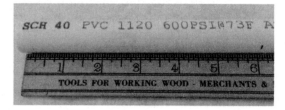

Figure 1-30 *Mark the PVC for a 6" cut length.*

Step 10

Measure down 1" from one end of the pipe, as shown in Figure 1-31.

Figure 1-31 *Mark 1" from one end of PVC.*

Step 11

Drill a 1/4" hole at the mark, as shown in Figure 1-32. Just drill through one side of the pipe.

Figure 1-32 *Drill a hole at the mark 1" from the end of the PVC.*

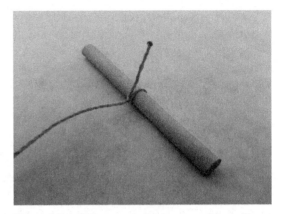

Figure 1-34 *Tie one end of string around the pull-handle.*

Step 12

Cut a 3' piece of nylon string and burn the ends to keep it from unraveling, as shown in Figure 1-33.

Step 14

Thread the string through the hole in the PVC pipe, then through the hole in the stick helicopter dowel, as shown in Figure 1-35. Drop the stick into the PVC and then wind the string around the dowel in the counterclockwise direction.

Figure 1-33 *Use a match to seal both ends of the nylon string.*

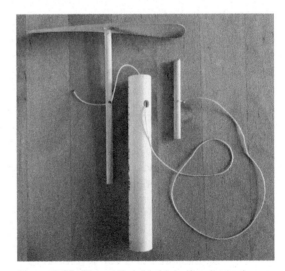

Figure 1-35 *Thread the string into the pipe and out the top then through dowel in the copter.*

Step 13

Tie a stick scrap onto one end of the string, as shown in Figure 1-34. This will be the pull-handle.

Step 15

While wearing safety glasses (to protect from rapidly moving blades), hold the PVC pipe firmly, as shown in Figure 1-36, raise it slightly above your head, and then give the cord a hefty pull. The copter should spin and rise out of the PVC pipe into the sky. If things don't work exactly that way, you might need to wind in the other direction or wind in such a way that the string releases more smoothly.

Figure 1-36 *Hold the PVC pipe tight and give the string a firm pull!*

Going Further

Below you'll find a short list of resources that will help you and your students gain further knowledge of helicopters and of how to modify the projects in this chapter.

The Rubber Band Helicopter by Workshop for Young Engineers (http://bit.ly/rubberband_heli)

> This project provides a lot of opportunity for hacking and trying new things with just a few materials and minimal cost.

Build Your Own Quadcopter (http://bit.ly/quad copter_drone)

> Love how helicopters work? *Make:* has a great project build for a remote control quad copter. You can build it using hardware store parts and R/C equipment that you can buy online. After that, you can add some sophisticated, but easy-to-use autopilot and camera hardware. With quads, you'll be jumping up significantly in price, but building your own can be hugely rewarding.

Stain PVC Any Color You Like (http://bit.ly/ stain_pvc)

> Check out this *Make:* project to make your pull-cord launcher look really cool.

Rockets are cool. There's no getting around that.

—Elon Musk, inventor

Ever since he was a kid, Rick has loved rockets. Solid propellant Estes rockets were the only rockets really available at the time, and Rick and his friends used to head out to the Mojave Desert for multistage launches that sent their rockets into the sky to altitudes well over 2,000 feet. The rocket launches were amazing, but with launches costing a few bucks a pop, they soon burned a hole in Rick's wallet.

Likewise, James also enjoyed launching rockets at a young age—the first rocket he built and launched on his own was a 15" tall rocket that was unfortunately lost on its fifth launch and never found again. Today, he teaches a summer camp where one of the week's projects is building and launching rockets on the school's baseball field. A portion of the camp's fee helps cover the cost of the one-launch-only engines—without the fee, the rocket launch party would have to be canceled.

Wouldn't it be great to be able to get the experience of launching rockets over and over again without having to spend money on disposable engines? With this project, we provide two rocket designs that use compressed air as propellant for free and clean launching every time!

After the rockets are built, the only costs involved for each launch may be tired muscles from using a bicycle pump. They are also safe to use in the city (and in dry scrubland) because there is no combustion, so the rockets pose no fire hazard.

The first project we present in this chapter is a simple $5 launcher that sends rockets 30–40 feet into the air. It's a good place to start, and very safe to use around young children. When you're ready to put some rockets a bit higher in the air, move on to the high-pressure launcher project also found in this chapter.

You may want to get a conversation on rockets started, so let us offer some questions for thought and discussion before we dive into the projects:

- Who is considered the inventor of rockets?

- What pushes a rocket to leave the atmosphere?

- Air rockets get all their thrust at once, in a split second. How do other types of rockets get their thrust?

- Try launching a rocket without the fins. What happens? Why are fins so important? Try a rocket with one, two, three, or four fins. Any difference in performance?

- Wanna get some math on? Try some launch calculations at NASA's "Air Rocket Launch Equations" page (*http://bit.ly/air_rocket_launch_eq*).

When working with anything involving projectiles, it's a good idea to wear safety glasses. The high-pressure rockets can be especially dangerous if you're not careful. Make sure the launch area is clear of people and pets that might get in the way of the rocket's blast off.

Low-Pressure Rocket Launcher

This project has two parts. First, you'll build a launcher, which is reusable and stays on the ground. Second, you'll build a reusable rocket that slides onto the launcher. The rocket will be launched into the sky using a the simple action of stomping on the launcher; this forces air through a tube and into the rocket, thus propelling the rocket upward.

This first launcher uses a low-pressure rocket, because even though a stomp can have a lot of energy behind it, it's still a one-time action that can only push a fixed amount of air into the rocket. One student might be able to get a few extra feet of altitude with a harder stomp than another student, but ultimately, most low-pressure rocket launches will result in the rocket going up 30–50 feet before coming back down. For this reason, launches are typically safe for smaller areas (e.g., a backyard or a playground) where the risk of losing the rocket is small.

For groups of 10 or more students, building additional launchers can help reduce the amount of waiting time to launch a rocket. Figure 2-1 shows the completed low-pressure launcher in action.

Build the Low-Pressure Rocket Launcher

The following steps include photos that can be used by the instructor or student to verify a task before moving on to the next step.

Step 1

Gather the supplies shown in Figure 2-2. You can find a complete list of the materials required for this project in "Tools and Supplies".

Figure 2-1   *Awesome low-pressure rocket launcher in action!*

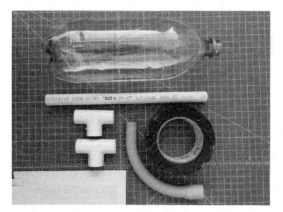

Figure 2-2 *The parts needed to build the low-pressure rocket launcher.*

Step 2

Cut the 8' piece of PVC into the following lengths: 3' (1x), 2' (1x), 1' (2x), 3" (1x). You can either use a hacksaw or a handy PVC cutter (Figure 2-3), available in the irrigation section of hardware stores for about $15. (If you use a hacksaw, wear a mask so that you don't breathe in the PVC dust!)

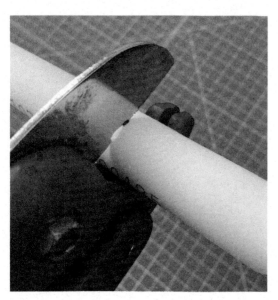

Figure 2-3 *A PVC cutter can cut through PVC quickly and cleanly*

Step 3

Connect the two tee fittings together with the 3" PVC tube, as shown in Figure 2-4.

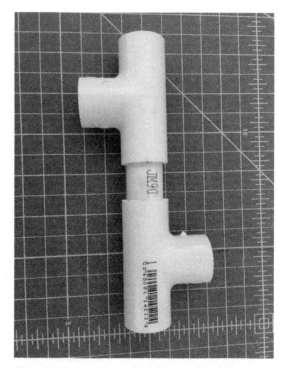

Figure 2-4 *Two PVC tee connectors joined with a single piece of PVC tube.*

Step 4

In the end of one of the 1' long pieces, add some hot glue inside the pipe and push tightly wadded paper into the end. Cover the wad of paper with hot glue to create an airtight seal, as shown in Figure 2-5. Do this to the ends of both 1' long pieces. (These two 1' pieces will form the legs of the launcher.)

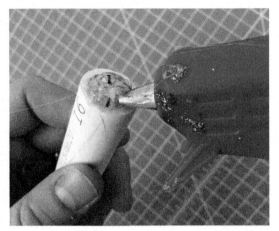

Figure 2-5 *Stuff in a wad of paper and then seal it with hot glue.*

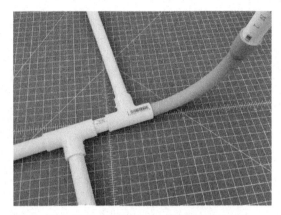

Figure 2-7 *Connect the 3' piece of PVC on one end of the stand and the 90 conduit elbow on the other.*

Step 5

Insert a plugged 1' pipe into one of the tee connectors, as shown in Figure 2-6. Insert the other plugged 1' pipe into the other tee connector.

Figure 2-6 *Insert a plugged end into a tee connector.*

Step 6

Put the gray PVC conduit 90 degree elbow into one of the tee connectors and insert the 3' PVC tube into the end of the other tee connector, as shown in Figure 2-7.

Step 7

Take the 2' long piece of PVC (this will be the vertical launch tube) and bevel one end using the concrete as a file. Angle the pipe at a 45 degree angle as you're rubbing it on the cement. This beveled end will be the top of the launch tube and will help the rockets slide on easier. Place the launch tube in the 90 degree gray conduit with the bevel at the top. You can see this 2' long piece inserted into the curved conduit in the upper-right corner of Figure 2-7.

Step 8

Slip the mouth of the 2 liter bottle into the 3' piece of PVC and seal with duct tape, as shown in Figure 2-8. The 2 liter bottle will be the pressure chamber. The completeed low-pressure launcher appears in Figure 2-9.

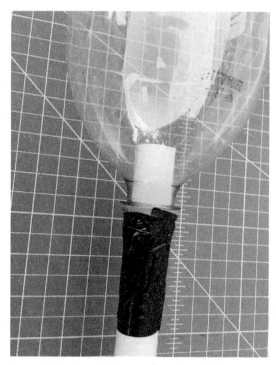

Figure 2-8 *Seal the bottle and PVC with duct tape and now you're done!*

Figure 2-9 *Completed low-pressure rocket launcher and paper rocket.*

Build the Low-Pressure Paper Rocket

Step 1

Collect all the materials shown in Figure 2-10: colorful paper (or white if you prefer), clear tape, plastic wrap, and a 12" piece of 1/2" PVC pipe, and pens for decorating.

Figure 2-10 *The supplies needed to build a rocket.*

Step 2

Wrap the paper around the remaining scrap piece of 1/2" PVC pipe and tape along the seam, as shown in Figure 2-11. Make sure you don't tape the paper to the pipe; the paper should slide off easily.

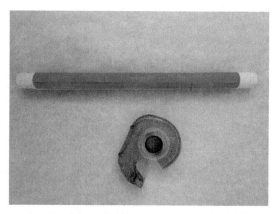

Figure 2-11 *Roll the paper around the PVC and tape closed.*

Step 3

Push the paper to the top of the pipe so about 1/2" of the paper hangs over the PVC, then fold it down, as shown in Figure 2-12.

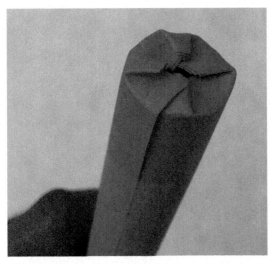

Figure 2-12 *Fold down a bit of the paper over the end of the PVC.*

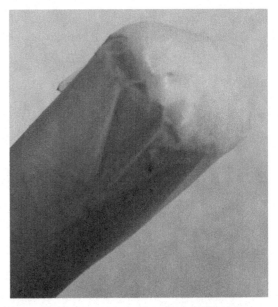

Figure 2-13 *Tape a piece of wadded plastic to the top of the rocket.*

Step 4

Crisscross two pieces of tape over the top to create a good seal. Wad up a small piece of plastic wrap and tape it securely to the top (using more tape), as shown in Figure 2-13. The plastic wrap makes a soft nose cone and also makes the rockets nose-heavy, so they'll arc over at the height of their trajectory.

Step 5

Cut four paper fins, as shown in Figure 2-14.

Step 6

Now we'll create tabs to tape the fins to the rockets. Stack the fins and fold them about 1/2" up on the long side. Cut a slit and then fold back in opposite direction, as shown in Figure 2-15.

Figure 2-14 *Cut four fins for your rocket.*

Figure 2-15 *Fold the fins along the long edge and cut a slit.*

Step 7

Tape all four fins (three fins will work, too) evenly around the bottom of the rocket, as shown in Figure 2-16, and you're ready to launch.

Step 8

To launch, place the rocket on the launch tube.

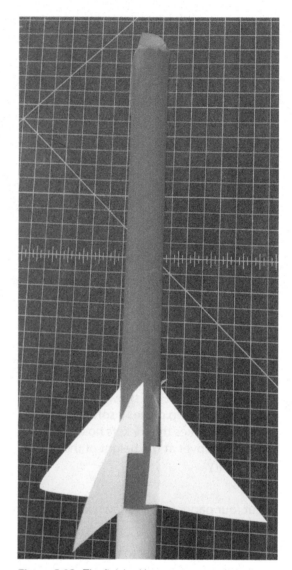

Figure 2-16 *The finished low-pressure rocket.*

Step 9

Stomp in the middle of 2 liter bottle instead of on the ends. This will help the bottle last 150–200 launches.

Step 10

After each launch, if you are not able to just squeeze the bottle back into shape, have the person who just launched wrap their hand around the launch tube (to minimize germs) and blow in to reinflate the 2 liter bottle pressure chamber.

High-Pressure Rocket Launcher for Super High Flights!

A few years ago, Rick came across a rocket powered strictly with compressed air that went about as high as the smallest motor-powered rockets that you ignite.

He worked on his own design for a compressed air rocket launcher, and presented it in *Make: Magazine* (Volume 15). The rockets are made of paper and tape, and blast hundreds of feet in the air, fueled only be a few pumps on a bicycle pump. Since then, thousands of launchers have been built around the world.

With the help of a friend, he recently refined the design, and it's been redeveloped as the project you're about to read. It can be built in about 30 minutes using the kit available at the MakerShed (*http://bit.ly/car_launcher_v2_1*) or AirRocketWorks.com. While the launcher kit costs just over $100, the rockets can be made for just pennies. The parts for the launcher are difficult to find and can be expensive, so the kit is really a great deal, especially if you'll be using it for a group of kids. The durable steel and wood industrial design will last for years of launching. We've used this same launcher at Maker Faires from New York to California (Figure 2-17), sometimes launching thousands of rockets over the course of a weekend.

Figure 2-17 *High-pressure rocket launcher in action at the National Maker Faire in Washington, DC! (photo: Kevin Jarrett)*

Build the High-Pressure Rocket Launcher

Step 1

Gather the supplies shown in Figure 2-18. You can find a complete list of the materials required for this project in "Tools and Supplies".

Figure 2-18 *Compressed Air Rocket Launcher v2.1 kit (available at MakerShed.com or AirRocketWorks.com).*

Step 2

Assemble the launch valve. Pull out the four parts you see in Figure 2-19 from the Compressed Air Rocket Launcher v2.1 kit. For the air to flow directly into and out of the launcher, it's important to assemble the launch valve in the correct direction. It's also very important that there are are no leaks in the system. If there are air leaks, the launcher will not work correctly. Add pipe thread (Teflon) tape to *all* screw-in connections.

Figure 2-19 *Parts placed in order for the launch valve*

Step 3

Start from the lefthand side of the image in Figure 2-19. Wrap plenty of Teflon tape around the 1/4" push-fit hose connection.

You need to make sure to apply the tape in the direction you'll be screwing it in, as shown in Figure 2-20. (You'll turn it clockwise to tighten, so wrap the Teflon tape in a clockwise direction as well.)

Figure 2-20 *Wrapping sealing tape (included in kit) on the 1/4" push-fit hose connector*

Step 4

Screw the push-fit connector into the "pressurize" side of the of the blue slide valve, hand tight, as shown in Figure 2-21. It's the side of the slide valve that has the hex end.

Figure 2-21 *Push-fit connector threaded hand-tight onto the hex ("pressurize") end of the blue slide valve.*

Step 5

Add Teflon tape to the check valve and hand tighten onto the "launch" side of the slide valve, as shown in Figure 2-22.

Figure 2-24 *The brass Schrader valve screws into the check valve.*

Step 8

Clamp the 1/4" push-fit connector on the "pressurize" side of the slide valve in a vise. Use a rag to keep it from getting scratched. Use an adjustable wrench on the Schrader valve to tighten all connections, as shown in Figure 2-25.

Figure 2-22 *Hand tighten the check valve, onto the "launch" side of the blue slide valve.*

Step 6

Add pipe thread tape to the 1/4" Schrader valve, as shown in Figure 2-23.

Figure 2-23 *Brass Schrader valve.*

Step 7

Hand tighten the Schrader valve into the check valve as shown in Figure 2-24.

Figure 2-25 *Tighten all connections at once using an adjustable wrench.*

Step 9

Now you will build the pressure chamber. Locate the 1/2" steel pipe cap and

1/2" × 12" steel pipe nipple from the kit, shown in Figure 2-26.

Figure 2-26 *1/2" steel cap and pipe nipple for pressure chamber.*

Step 10

Put a generous amount of Teflon tape on one threaded end of the 12" pipe and thread the 1/2" cap on hand tight, as shown in Figure 2-27.

Figure 2-27 *1/2" cap hand tightened onto 12" pipe nipple.*

Step 11

Add Teflon tape to the other end of the 12" steel pipe and thread it into the "A" side of the QEV valve. The QEV is the piece marked A, R, and P shown in Figure 2-29. Use a rag to protect the 1/2" cap and hold the cap with a pair of pliers, as shown in Figure 2-28, and then tighten both connections with an adjustable wrench on the QEV side. (You can also clamp the pipe cap in a vise vertically and then tighten with the adjustable wrench.) *Note: The QEV valve is*

made out of aluminum, so make sure you don't strip the threads by over- tightening.

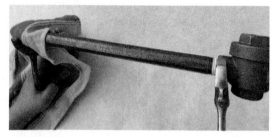

Figure 2-28 *Tighten connections on the pressure chamber.*

Step 12

Figure 2-29 shows the QEV "P" side where you will screw in the 1/2"-to-1/4" reducer first, followed by the second 1/4" push-fit hose connector. Wrap both the reducer and the push-fit hose connector with pipe sealing tape before making the connections. With your adjustable wrench on the push-fit connector, tighten both connections—but don't over-tighten!

Step 13

Now locate the remaining parts from the kit, along with the assembled slide valve and pressure chamber, as shown in Figure 2-30.

Step 14

Feed the 1/2" cap side of the pressure chamber through the keyhole in the small piece of the wood stand, as shown in Figure 2-31. Feed it about halfway through and then pop it into the notch, as shown in Figure 2-32.

Figure 2-29 *With sealing tape on all threads, tighten the connectors shown here to the P side of the QEV.*

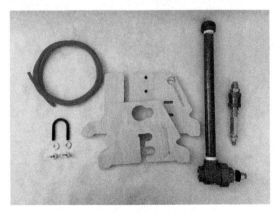

Figure 2-30 *Locate the pieces shown here from the kit to complete the stand and finish the launcher.*

Figure 2-31 *Fit 1/2" cap on the pressure chamber into the keyhole in the small wood stand piece.*

Step 15

Now fit the small wood stand piece into the large piece, as shown in Figure 2-32. If you're not sure if it's facing correctly, look at the completed photo in Figure 2-39. The stand is designed to fit snugly. Make sure everything is in the right direction, and then pound the two pieces together with your hand. If more force is needed, use a rubber mallet and piece of scrap wood on top to keep from damaging the stand.

Figure 2-32 *Fit stand together and pound gently with a rubber mallet to fit together (check Figure 2-39 for correct orientation).*

Step 16

Once the stand is firmly together, slide the QEV down until it presses against the edge of the wood stand on the right side, as shown in Figure 2-33.

Step 17

Put the U-bolt through the two holes around the 1/2" pipe, as shown in Figure 2-34.

Step 18

On the other side, slip the washers on and tighten down with the wing nuts. You can see this in Figure 2-35.

Figure 2-33 *Slide QEV against the end of the wood stand.*

Figure 2-34 *Slip U-bolt around the steel pipe and through the holes in the wood stand.*

Step 19

Grab the red hose and insert one end firmly into the 1/4" push-fit connector on the slide valve launch assembly, as shown in Figure 2-36.

Figure 2-35 *Slip washers on and tighten the wing nuts.*

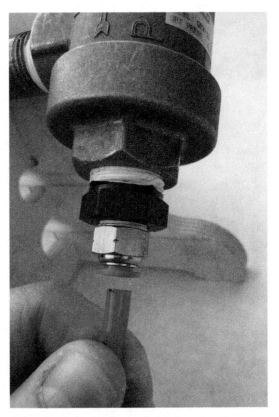

Figure 2-37 *Push other end of air hose into the push-fit connector on the "P" side of the QEV.*

Step 21

Screw the 1/2" × 12" gray PVC tube onto the "R" side of the QEV. This is the only connector that does not require pipe seal tape, as it is not under pressure. You can see this in Figure 2-38.

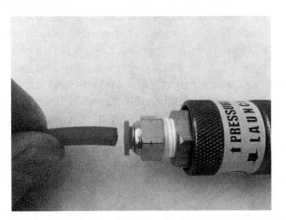

Figure 2-36 *Push red air tube firmly into the push-fit connector in the slide valve.*

Step 20

Push the other end of the hose firmly into the "P" side of the QEV valve, as shown in Figure 2-37.

Figure 2-38 *No pipe seal tape necessary.*

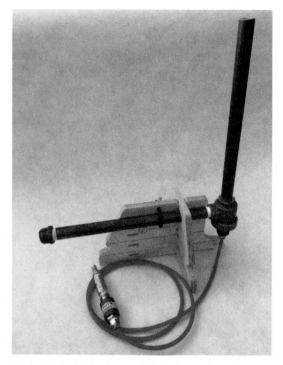

Figure 2-39 *Completed high-pressure rocket launcher.*

Step 22

Your launcher is now complete and should look like the one in Figure 2-39! In order to tilt the launch tube, just loosen the wing nut and pivot the pressure chamber, and then retighten.

Storing the High-Pressure Rocket Launcher

To store your launcher, unscrew the gray PVC launch tube, and insert it horizontally into the bottom hole of the wood stand. The middle hole is for the 3/8" launch tube which is used to launch other fun things. You can also cut two 4 1/2" pieces off the Velcro strip included in the kit. Secure both tubes with the long Velcro strap going through the slots in the launcher. Wrap the hose around the launch stand and secure with the remaining 2" piece of Velcro, as shown in Figure 2-40.

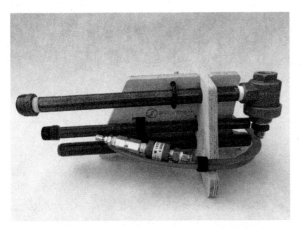

Figure 2-40 *High-pressure launcher folded for storage.*

Now that you have your launcher, it's time to build a rocket (or two...or two dozen) that you can actually launch. Fortunately, the steps for making a rocket are super simple. Once you make a few, you'll discover that you can make many rockets in a short amount of time. To make rocket assembly easier, there are directions for a handy rocket building stand in the "Rocket Stands" project. After building your rockets, skip to the "Launching Rockets!" section to see how to use your launcher.

Build the High-Pressure Rockets

Step 1

Gather the materials shown in Figure 2-41: the printed rocket template (*http://bit.ly/ mpgpr_templates*), the 1/2" × 12" grey PVC pipe, scissors, and 3/4" wide masking tape.

Step 2

Cut out the body tube, pressure cap, nose cone, and three fins, as seen in Figure 2-42.

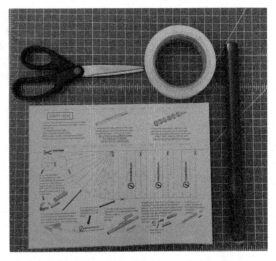

Figure 2-41 *Materials and tools needed for the paper and tape rocket.*

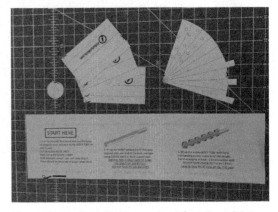

Figure 2-42 *Rocket pieces cut out from template.*

Step 3

Wrap the body tube tightly around the gray PVC pipe and tape in the middle, top, and bottom, as shown in Figure 2-43. *Do not* tape the paper body tube to the PVC pipe: it should slide on and off the pipe easily. The PVC pipe is just used for building the rocket. While taping, make sure you keep things as smooth as possible for a sleek, aerodynamic rocket that will go really high!

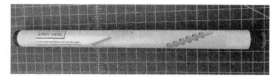

Figure 2-43 *Body tube wrapped and taped around gray PVC build tube (do not tape the paper to the PVC pipe; it should slide off).*

Step 4

Now, working your way down the body tube, overlap strips of masking tape, as shown in Figure 2-44, making sure no paper is exposed (otherwise it may blow out during the launch).

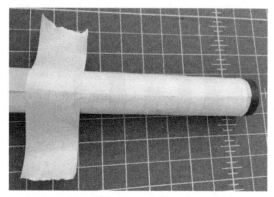

Figure 2-44 *Work your way down the body tube, overlapping tape along the way.*

Step 5

Slide the body tube up to to the top of the PVC pipe and place the pressure cap on top, as shown in Figure 2-45.

Step 6

Take two pieces of tape and crisscross them over the pressure cap and smooth down, as shown in Figure 2-46. This is where all the pressure at launch goes, so it's important to get a tight, strong seal. Wrap one more piece of masking tape around the body tube at the top.

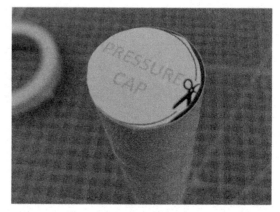

Figure 2-45 *Place pressure cap on top of tube.*

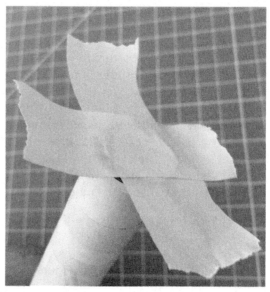

Figure 2-46 *Crisscross tape over the top of the pressure cap.*

Step 7

Take the nose cone piece and pull it along the edge of a table to curl the paper in the direction you want. Overlap the nose cone into a cone shape over the dots on the template. Pack the nose cone tightly with scraps of paper using your scissors to push the paper down, as shown in Figure 2-47.

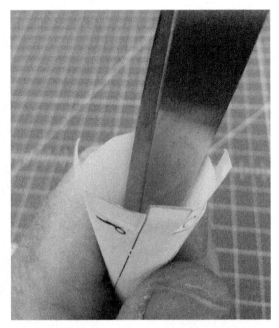

Figure 2-47 *Stuff nose cone tightly with scraps of paper*

Step 8

Tape the nose cone over the pressure cap using the tabs to the top of sealed end of the body tube, as shown in Figure 2-48. Cover the entire nose cone in strips of tape.

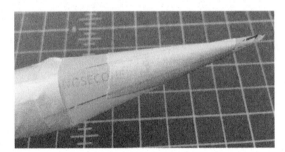

Figure 2-48 *Tape nose cone to the top of the body tube and then cover nose cone with strips of tape.*

Step 9

Fold the three fins along the center line, as shown in Figure 2-49.

Figure 2-49 *It's easier and more uniform to fold all three wings together.*

Step 10

Fold the left and right edges of the three fins back (in the opposite direction of the center line fold). These are tabs you'll maybe use to tape the fins to your rocket. You'll be left with three fins that look like the ones in Figure 2-50.

Figure 2-50 *Three fins with tabs for taping to your rocket.*

Step 11

Use masking tape to evenly space the three fins around the bottom of the rocket. One piece of masking tape can hold the edge on two fins, as shown in Figure 2-51.

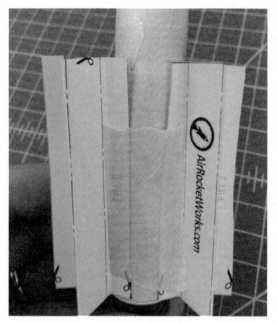

Figure 2-51 *One piece of tape can secure the tabs on two fins.*

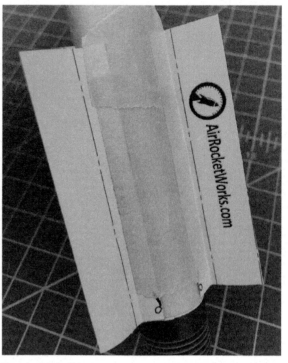

Figure 2-52 *Tape the fins all the way up and down so they don't tear off at launch.*

Step 12

Once the fins are secure, use additional masking tape to completely cover the folded edges, as shown in Figure 2-52.

A finished rocket is shown in Figure 2-53.

Figure 2-53 *Rocket ready for launch!*

Launching Rockets!

If you're able, it's best to place your launcher up on a table so the launch tube is above eye level. Always make sure everyone is clear of the launcher before launching. Having an adult in control of the launch valve is always a good idea.

Step 1

Make sure the slide valve is in the "pressurize" position (Figure 2-54). Connect your bicycle pump to the Schrader connector side of the slide valve (Figure 2-55). Make sure you have a good, airtight connection to the pump. A high-quality pump we recommend is the Lezyne Alloy Floor Pump.

Step 2

Place your paper and tape rocket on the launcher (Figure 2-56).

Figure 2-56 *Launcher connected to bike pump and ready to go.*

Figure 2-54 *Slide valve in the "pressurize" position.*

Figure 2-55 *Connect bike pump to the brass Schrader valve (right side of picture) securely; make sure there are no leaks.*

Step 3

Using a pump with a pressure gauge, pump it up to about 50 psi like in Figure 2-57. It should only take about 2–3 pumps.

Step 4

With everyone standing clear of the launcher, count down from 5 and then slide the valve back to the launch position, as shown in Figure 2-58. All the air should instantly be released from the pressure chamber and the rocket should fly nearly 250′ (almost out of sight), before falling back to earth.

Figure 2-57 *Pump up to 50 psi and pressure should hold if there are no leaks in the system.*

Figure 2-58 *Slide valve toward the "launch" position to fire the rocket!*

Step 5

After launching, slide it back to the "pressurize" position only *once*. This keeps pressure in the line and helps the QEV reset itself. If the QEV does not reset, you may need to work the slide valve back and forth to reset the plastic diaphragm inside the QEV. You'll hear a faint "click" when it resets. Further troubleshooting tips can be found at AirRocketWorks.com. It's also possible to connect the launcher to an air compressor if you're doing hundreds of launches like we do at Maker Faire. See AirRocketWorks.com for tips on how to do that as well.

Going Further

Experiment!

For both the low-pressure and high-pressure rockets, try building rockets out of different types of paper and tape. Make the fins different shapes and try a variety of materials. See how the changes affect the flight performance.

Several more things that can be launched with the high- and low-pressure launchers are found in Chapter 4. There are lots of opportunities to modify and experiment with parachutes, rocket gliders, and roto-recovery.

The rocket and launcher design are being continually improved. For the latest updates, go to AirRocketWorks.com.

Airplanes and Gliders 3

The airplane stays up because it doesn't have the time to fall.

—Orville Wright

The Wright brothers are a classic example of the trial and error that inventors have experienced throughout history. After numerous attempts, involving kites and gliders of varying sizes and developing their own gasoline engine, on December 17, 1903, the brothers finally achieved the first successful powered flight —a distance of 120 feet in 12 seconds. By the end of the day, they had flown 852 feet and stayed in the air a grand total of 59 seconds. They finally got the right combination of power, lift, and control. Within 20 years of that historic flight, air travel had become a part of everyday life.

An airplane has four forces working on it: thrust, lift, drag, and weight. Thrust is the force that moves the plane forward, provided by the propeller. Lift is a force cause by a low-pressure system forming over the wing while the plane is moving forward. Drag is the force *against* the plane's movement, caused by the resistance of the air (much like when you put your hand out the window of a car). Weight is the force of gravity, trying to pull the plane back down to the earth. To combat drag and weight, planes are made as lightweight and sleek as possible to help the thrust and lift forces move the plane through the air.

For this next project, we've got a fun little rubber band airplane that will allow you to experiment in many ways to get ideal flights. A well-built airplane with the right weather conditions can easily fly for over 30 seconds. Following the rubber band airplane project, you'll also find the first of its kind air rocket glider (which we'll launch from the high-pressure compressed air launcher we made in Chapter 2).

Here are some suggestions for discussion questions:

- Why do airplanes stay up in the air so much more easily than helicopters?
- Why can airplanes have much smaller motors than helicopters?
- How do gliders stay up in the air without a propeller?
- Do airplane wings work in a similar manner to a bird's wings?

Rubber Band Airplane

Airplanes come in all sizes (and in some unusual shapes, as well), from the large passenger planes that can carry hundreds of people over land and sea, to the single-seaters that provide a nice arial view to both the passenger and pilot.

For this project, however, you won't be making an airplane that can ferry any passengers or cargo. You will, however, be making an airplane that behaves just as its bigger siblings do in the air—flying via lift provided by thrust, and fighting both drag and gravity to stay in the air. You can see exactly what this little flying machine looks like in Figure 3-1.

Figure 3-2 *Rubber band airplane supplies.*

Step 2

Print the templates that are included at *http://bit.ly/mpgpr_templates*. On one of the three plates, lay down the smaller rudder and elevator templates and then trace and cut out, as shown in Figure 3-3. Save the foam scraps, as you'll be using them later.

Figure 3-1 *This is the airplane we'll be building with soda straws and foam plates—it's super light and flies awesome!*

Build the Rubber Band Airplane

The following steps include photos that can be used by the instructor or student to verify a task before moving on to the next step.

Step 1

Go ahead and gather the materials shown in Figure 3-2. You can find a complete list of the materials required for this project in "Tools and Supplies".

Figure 3-3 *Trace and cut out the rudder and elevator components.*

Step 3

Lay down the 3.5" × 5" wing template in the center of a second plate. Trace and cut around three sides as shown in Figures 3-4 and 3-5.

Figure 3-4 *Trace around the left edges and bottom edge.*

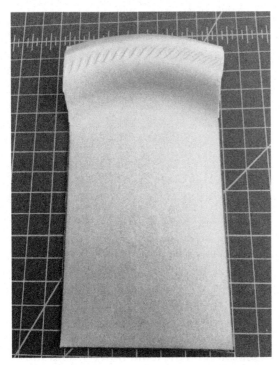

Figure 3-5 *Cut along the lines and include the curved portion of plate.*

Step 4

Trim the curved end of the wing into a nice, evenly curved shape, as shown in Figure 3-6. Use this wing as a matching template to make the other side of the wing, as shown in Figure 3-7.

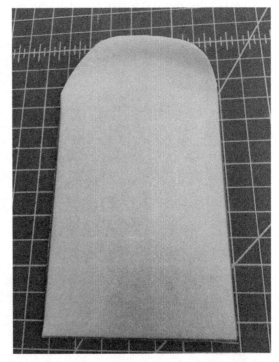

Figure 3-6 *One of the two wings.*

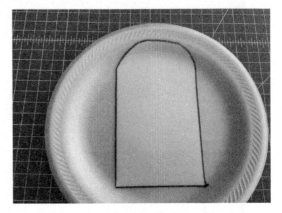

Figure 3-7 *Trace the cut out wing to make a duplicate wing.*

Step 5

Now that you have two parts to the main wings, you can attach them together. Join both ends together and then tape together using clear tape, as shown in Figure 3-8. Flip the main wing over and glue a sliver of scrap foam over the top of the seam for a super strong bond (Figure 3-9).

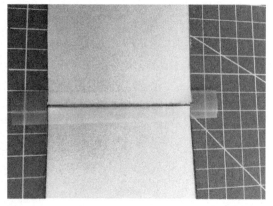

Figure 3-8 *Tape both wings together.*

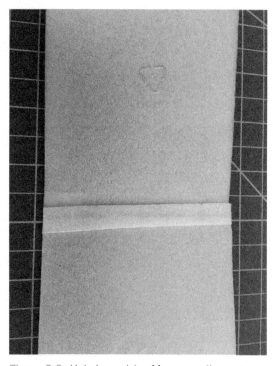

Figure 3-9 *Hot glue a strip of foam over the seam to create good bond between wing halves.*

Step 6

You'll next create a slight curve in the wing by cutting three 1" scraps of foam, as shown in Figure 3-10. Then stack and glue the three scraps onto the center of the main wing, as in Figure 3-11.

Figure 3-10 *Mark and cut three scraps of foam about 1" long and 1/2" wide.*

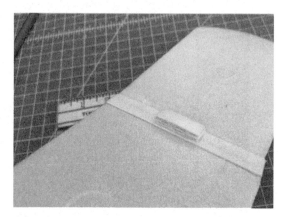

Figure 3-11 *Hot glue them in a stack onto the center line of the wings.*

Step 7

Pinch the end of one straw and insert it as far as you can into the end of the other straw, as shown in Figure 3-12. Wrap clear tape around the joint. Trim the long straw tube to an 11" length. The straw will be your motor stick.

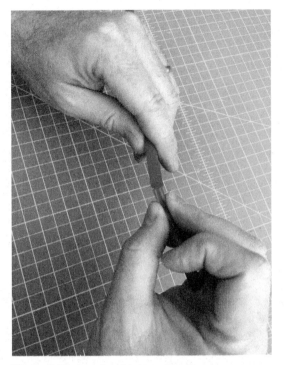

Figure 3-12 *Insert soda straws together.*

Step 8

Mark the center of the horizontal stabilizer, line up with the rear of the straw, and glue in place with hot glue. The shorter side should be at the rear of the airplane (Figure 3-13). Do the same with the vertical stabilizer.

Figure 3-13 *Stabilizers.*

Step 9

From the front of the straw motor stick, mark 2 1/4". (Figure 3-14). Then flip the airplane over and glue the main wing onto the straw motor stick with the leading edge of the main wing lined up with the mark (Figure 3-15). Make sure everything is lined up straight!

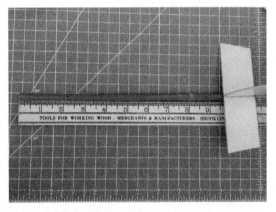

Figure 3-14 *Mark the front.*

Figure 3-15 *Glue the wing to the straw motor stick.*

Step 10

Now we'll connect the power system. Cut a small paper clip, as shown in Figure 3-16. Then cut a piece of rubber band to 28" and knot together, as shown in Figure 3-17.

Figure 3-16 *Cut a paper clip.*

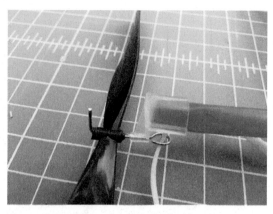

Figure 3-18 *Rubber band on the hook.*

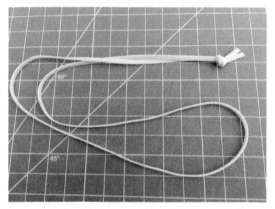

Figure 3-17 *Knot a rubber band.*

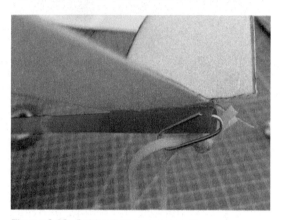

Figure 3-19 *Rubber band on paper clip.*

Step 11

Pinch the front of the straw motor stick and slip the propeller over the end. Slip the rubber band into the metal hook on one end (Figure 3-18), then onto the paper clip on the other (Figure 3-19).

Step 12

Now we'll make adjustments for good flights. Four things can be adjusted for nice and level flights: center of gravity, wing camber, angle of attack and wing dihedral. If your plane stalls, add a tiny bit of weight to the front. If it dives too quickly, add a bit of weight to the rear of the plane (Figure 3-20).

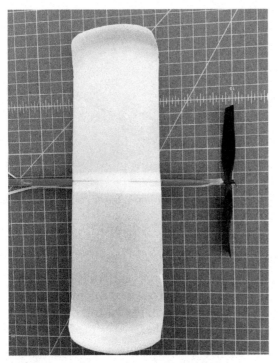

Figure 3-20 *Adjust the weight balance.*

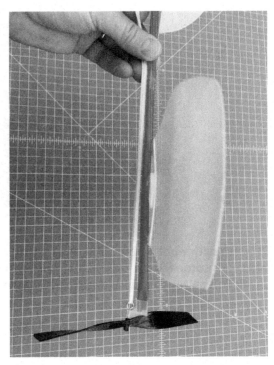

Figure 3-21 *The cambered wing.*

Step 13

Wing camber is the curve to the wing. The cambered wing creates lift (Figure 3-21). The angle of attack is the angle of the wing in relation to the fuselage (in our case, the soda straw). This can be adjusted by raising or lowering the leading edge of the main wing. Finally, the dihedral is the distance from the tip of the wing down to the work surface. See Figure 3-41 in the next section for a good diagram showing both angle of attack and dihedral. You might have to unglue the wing, and move and reglue to get it into the position you want. It's possible the airplane will fly great right away!

Step 14

Flying! A battery powered winder is handy because once you have the airplane tuned correctly, you'll want to wind nearly 1,000 times! (Figure 3-22) Once wound, hold plane level, release the propeller for a burst of power and give a gentle toss. Half the fun of flying is continuing to adjust the plane until you get awesome flights. Try adding paper flaps to the trailing edge of the main wing to get the airplane to climb steeply.

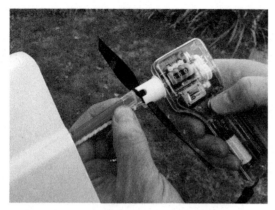

Figure 3-22 *Use a battery powered winder.*

Figure 3-24 *CAD drawing of the air rocket glider (ARG).*

If you really want to get ambitious and make your own propeller as well, Slater Harrison has some helpful instructions (*http://www.science toymaker.org/plane/propeller.htm*).

Air Rocket Glider Kit

Back in 2008, Rick designed the compressed air rocket launcher for *Make: Magazine*. Since then, it's been developed into a kit (which we used in "High-Pressure Rocket Launcher for Super High Flights!") and thousands have been built by individuals and groups all over the world. Later, Rick wrote up a step-by-step build on a balsa folding-wing glider developed by Jim Walker with a history dating back to the 1940s. The glider rockets up with a rubber band catapult and at its height, the wings pop open and it glides down. (See "Going Further" at the end of this chapter for a web link to directions on building this balsa glider.)

Rick always dreamed of combining this magical glider with an air rocket, and then he met Keith Violette, who had been working on a prototype. Presented here is a step-by-step build of the glider kit Keith developed (Figure 3-23). The first of its kind air rocket glider (Figure 3-24) launches using the high-pressure compressed air rocket launcher (from Chapter 2). The kit can be found at AirRocketWorks.com.

Build the Air Rocket Glider

Step 1

Gather the materials and tools shown in Figure 3-25. You can find a complete list of the materials required for this project in "Tools and Supplies".

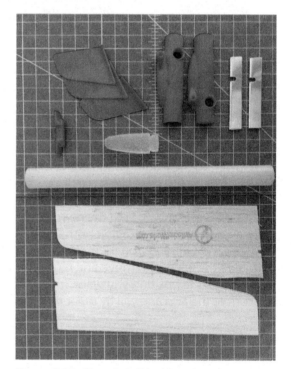

Figure 3-25 *Air rocket glider kit parts.*

Step 2

Print out the three fin guide (*http://bit.ly/ mpgpr_templates*). Mark where the three

fins should be placed, as shown in
Figure 3-26.

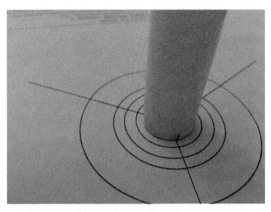

Figure 3-26 *Use the guide to mark where the fins will be placed on the tube.*

Figure 3-23 *With wings folded back, the air rocket glider launches from the high-pressure air launcher; the wings snap open at apogee for a gentle glide down. (photo: Jeffery Braverman)*

Step 3

Wrap a piece of masking tape a distance of 2.5" from the bottom of the tube where you made the marks for the three fins, as shown in Figure 3-27. Use a piece of fine sandpaper to rough up the portion of the tube below the tape so the fins stick better when you glue them on in a later step.

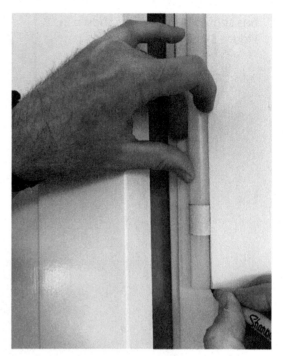

Figure 3-28 *Finish lines from the marks to the masking tape; mark one line the full length of the body tube.*

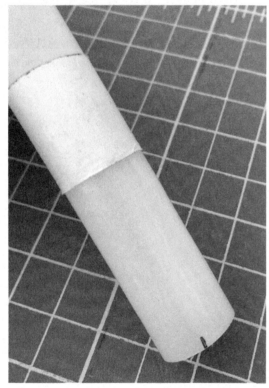

Figure 3-27 *Rough up the area below the tape with sandpaper.*

Step 4

Using a doorjamb to get a straight line, finish the marks made in step 3 around the bottom of the tube. Draw the lines all the way up to the masking tape, as shown in Figure 3-28. For one of the lines, draw the entire length of the body tube.

Step 5

Next, we're going to assemble the fuselage. Place the two plastic alignment pins in the two holes and rubber nose into the nose pocket in one side of the fuselage (Figure 3-29). Test fit it together before adding glue. Both halves should seat together perfectly. If they don't, clip the alignment pin a bit. Once you're sure you have a good fit, add super glue, and then clamp and let cure for a few minutes. If glue gets in the pivot hole, clean out with a bit of sandpaper (Figure 3-30).

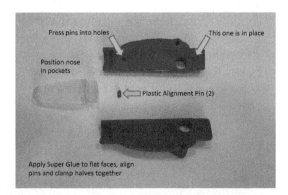

Figure 3-29 *Begin assembly of the main body.*

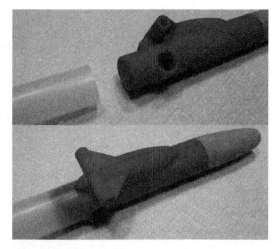

Figure 3-31 *Align fuselage on line running up main body.*

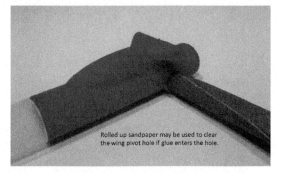

Figure 3-30 *Use sandpaper to clean out hole.*

Step 6

Attach fuselage to the body tube and glue on fins. Test fit the fuselage into the plastic body tube. Add super glue and line up the top center of fuselage with the line you drew all the way up the body tube. (Figure 3-31). Glue fins on as shown in Figure 3-32.

Figure 3-32 *Fins attached to tube body.*

Step 7

Slide the aluminum wing clip onto the edge of the wing so that the notch lines up and there is room for the wire to slide between the wing clip and wing (Figure 3-33). Print out the wing template form (*http://bit.ly/ mpgpr_templates*) and insert a staple through the bottom (side without the logo) of the balsa wood wing at the designated spot (Figure 3-34).

Figure 3-33 *Slide aluminum wing clip onto wing.*

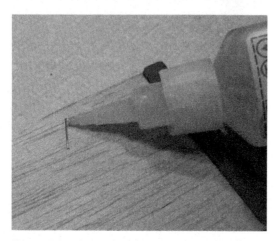

Figure 3-34 *Insert staple into balsa and secure with drop of glue.*

Step 8

Next, bend the pivot wire. Starting halfway down the length of the 9" wire, using needle-nose pliers, form a gentle radius in the wire that matches the outer diameter of the body tube (Figure 3-35). The legs of the wire should be parallel to one another, and

roughly equal in length—they don't have to be perfect, they will be trimmed later. As shown in Figure 3-36, grip the wire just below the midline of the tube. Make an approximate 100° bend in each leg of the wire.

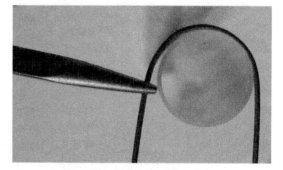

Figure 3-35 *Curve wire to match shape of tube body.*

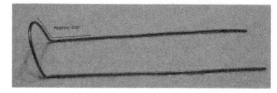

Figure 3-36 *Create two legs by bending wire to an angle of approximately 100 degrees.*

Step 9

Insert the rubber band through the upper tube-like hole in the fuselage. If needed, a short length of small gage wire or a small paper clip can be bent in a "U" and used to thread the rubber band through the tube. (Figures 3-37 and 3-38).

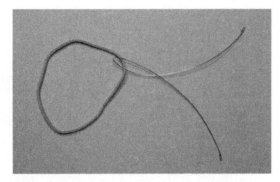

Figure 3-37 *Use thin wire to thread rubber band.*

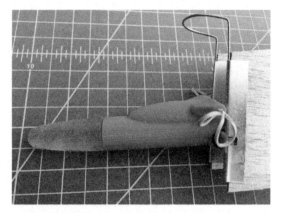

Figure 3-39 *Attach wings (with aluminum brackets) to the wing pivot.*

Step 11

Balance and fly! Stretch the rubber band and hook each end of the band to the formed staple hook on each wing. Be careful, as you can easily pull the rubber band out one side of the body. You should now be able to test the folding action of the wings. When released, the wings will hinge forward on the pivot wire, and then rotate on the plastic pivot into a gliding position as shown in Figure 3-40. Ensure they open quickly, evenly, and smoothly. If one side opens faster than the other, equalize the tension in the rubber band between the fuselage and the staple on each wing. When you "pluck" the rubber band like a guitar string, the sound should be the same on each side, indicating equal tension.

Figure 3-38 *Thread rubber band through plastic tube.*

Step 10

Putting it all together! This is a challenging step and involves a bit of patience, as six hands are needed. Just relax, and it will eventually come together. See Figure 3-39 to see how everything lines up. Fit the plastic wing pivot into the fuselage and align with the notches in the aluminum clips on the wings. Ensure the hooks formed by the staples are facing away from the body of the plane. Slide the bent wing pivot wire along the base of each wing, starting at the rear of the wings. It should slide along the base edge of the wings, through the holes in the wing pivot, and out the leading edge of the wing. It may take a couple tries to get it seated, and sometimes it helps to have an assistant. Pivot the wings to ensure smooth operation. Mark the excess wire flush to the front edge of the wings, and trim to length.

Figure 3-40 *How wings expand on rocket body.*

Step 12

Check the angles of the wings in the deployed position shown in Figure 3-41. The angle of attack can be adjusted by altering the two 100° bends in the wing pivot wire. The dihedral angle should be correct as supplied, but can be altered by adding tape or thin shims to the top of the wing where the wing pivot contacts the wing reinforcement at the base of each wing. Greater dihedral angle makes the plane more steady, but reduces lift (3° to 6° usually works well for this plane).

Figure 3-41 *Adjust wings for proper flight.*

Step 13

Due to varying densities throughout balsa wood, it is important that the ARG be balanced left to right. To do this, simply invert the plane and balance it so it can roll side to side on your fingers. If the plane always rotates so one wing is lower than the other, that lower wing is slightly heavier than the other. Pieces of tape can be added to the tip of the lighter wing tip, until the plane balances evenly, as shown in Figure 3-42. This will help the plane fly straight and true. If you are flying your ARG in a smaller field or park, you can purposely weight one wing tip to intentionally upset this balance. This will cause the ARG to spiral down to the ground, and not drift too far from the launch site.

Figure 3-42 *Completed air rocket glider.*

Step 14

The ARG launches off of a 3/8" NPT pipe, about 12" long, which is threaded on one end to connect to the launcher valve. There is also a bent piece of wire that holds the wings in the folded position until the ARG is launched. The new Version 2.0 and 2.1 Compressed Air Rocket Launcher kit (Figure 3-43) includes this launch tube, adapter, and the piece of wire for the wings. Directions for launching can be found at the end of Chapter 2, following the same instructions for launching high-pressure rockets.

Figure 3-43 *Air rocket glider on launcher ready to go.*

Step 15

If the ARG is to be used on a windier day, a second rubber band can be added to increase the opening power of the wings. This will cause the wings to deploy slightly sooner at a lower altitude, but will help prevent the wind from causing the plane to tumble or spin without opening its wings fully. Also, as the rubber band gets old and tired, it should be replaced to ensure proper wing operation. For storage, it is helpful to unhook the rubber band from the wings to prevent it from stretching.

The "Air Rocket Glider Launch Test" video (http://bit.ly/ arg_launch_test) shows the air rocket glider in action.

Going Further

The Balsa Folding-Wing (Rocket) Glider (http:// bit.ly/folding-wing_glider)

This is the project that the air rocket glider is based on.

"The Towel" R/C Stunt Plane (http://bit.ly/ the_towel)

After many years of trying, I learned how to fly R/C with the entirely homebuilt FLACK. Breck Baldwin at the Brooklyn Aerodrome is the designer of this cleverly simple airplane. You'll crash it a lot of times, but keep getting it up in the air again and again until you've gotten the hang of it.

Make: Hanger Tutorials with Lukas Weakley (http://makezine.com/makerhangar/)

This is a one-stop website for all the basics on R/C aircraft.

Ornithopter (http://bit.ly/build_ornithopter)

Also on Make: Projects is a rather complex but fascinating tutorial on building your own flapping, bird-like flying machine.

The Fantastic Foam Flier (http://bit.ly/foam_flyer)

Bill Kuhl, who is the brains behind the foam plate airplane, has shared this brand-new design, which is super stable and utilizes balsa and foam in its construction.

Rockets Revisited

4.

There's no doubt that building and launching rockets is a favorite activity of many kids and adults. For that reason, we've included some bonus rocket projects in this chapter that will really give you a variety of rockets and more hands-on rocket time.

We recommend that students start with the paper rockets back in Chapter 2. First, they're easier to build. Second, students will likely discover weaknesses in those types of rockets that will start them wondering how they might improve the design. That's when you'll want to introduce the first project in this chapter: the foam air rocket. More durable and capable of different flight styles, it's the obvious project to follow-up with after the paper rockets are done.

You will also find instructions for building a rocket-building base—perfect for building lots and lots of paper and tape rockets. It will definitely help speed things up in a classroom if you have a few of these or a dozen. We use these at Maker Faire to build thousands of rockets over the course of a weekend.

Finally, we'll give you two new rocket designs—one will use a super-bouncy ball in place of a nose cone and will give a nice big bounce when the rocket comes down after launch. The other will drift gently down after a packed parachute opens up—just make sure you don't launch this one on a really windy day or you could be chasing a rocket for some distance!

And here are some additional questions to open up further discussion on the topics of rockets:

- Most rockets are launched by countries/governments, but not all. Can you identify any private companies that launch rockets? (Hint: Check out *http://www.spacex.com*)

- In the United States, we call those who go into space *astronauts*. What other names are used (in other countries) to describe humans who go into space?

- Rockets are often launched and divided into sections as they get higher and higher. These are called stages, and rockets are often referred to as multistage rockets. Investigate the reasons for launching people and satellites into space using multistages.

- Most rockets use some form of combustible material to fire into space. What are some other methods used for sending rockets into orbit?

- The Space Shuttle program was revolutionary because much of the equipment was reusable. What modern-day space programs use reusable equipment?

Foam Air Rocket

Back in Chapter 2, you created a simple rocket using nothing but paper and tape. The low-pressure rocket probably got up to about 30 or 40 feet before coming back down. Not bad, but how would you like to make a rocket that goes higher and is still just as safe to launch? Take a look at Figure 4-1 and you'll see a slight variation of the low-pressure rocket called the foam air rocket. This one is made from foam and duct tape, but still uses the compressed air launcher made in "High-Pressure Rocket Launcher for Super High Flights!".

Build the Foam Air Rocket

Step 1

Go ahead and gather the materials shown in Figure 4-2. You can find a complete list of the materials required for this project in "Tools and Supplies".

Step 2

Cut a 9" section of foam pipe insulation. Wrap a single zip tie 1" from one end and cinch it tight so no air will escape, as shown in Figure 4-3. Clip off the excess zip tie with wire cutters, then trim the excess foam above the cinched-off end.

Figure 4-1 *Foam and duct tape will make this rocket more durable.*

Figure 4-2 *Materials needed for the foam air rocket.*

Figure 4-3 *Cinch zip tie 1" from end of foam piece.*

Step 3

Cut off a 4" length of duct tape and tear it lengthwise and crisscross these two pieces of duct tape over the cinched end. You can use another piece of duct tape wrapped around the foam body to hold these two crisscrossed pieces in place. You can see a piece of tape wrapped around the criss-crossed pieces in Figure 4-4. Next, cover the remaining visible foam section with duct tape spanning the length. Be especially sure you thoroughly cover the slit that runs the

length of the foam, so that the foam doesn't blow out at launch.

Figure 4-4 *Cover entire foam piece with duct tape.*

Step 4

Cut a 4" × 11/2" rectangle from the foam sheet, then cut it diagonally to make two fins. Repeat this step to create a third fin; the fourth fin can be saved for another rocket. Figure 4-5 shows fins cut in various colors.

Step 5

In order for the fins to stick securely to the foam rocket, you need to wrap a smooth layer of clear tape around the bottom of the foam rocket body, as shown in Figure 4-6.

Step 6

Using generous amounts of hot glue, attach the three fins, spaced evenly, onto the foam rocket body tube, as shown in Figures 4-7 and 4-8.

Figure 4-5 *Foam fins cut and ready to be glued on.*

Step 7

Pressurize the compressed air launcher for 45psi–65psi, and launch. (For a lower pressure/altitude launch, use a stomp launcher.) When the duct tape finally fails with a spectacular blowout, just apply more duct tape over the blown section and keep flying!

Figure 4-6 *Add a 4" wide layer of clear tape around the bottom of the rocket.*

Figure 4-7 *Apply hot glue along the long edge of each fin.*

Figure 4-8 *Three fins hot glued to the bottom of the foam rocket.*

Rocket Stands

Back in Chapter 2, you learned how to make a single low-pressure rocket from paper and tape. It's not hard to make a single rocket, but if you've got a lot of rockets to make with big groups like we do every year, rocket-building stands like the one seen in Figure 4-9 are a cheap and handy thing to use. We use our stands for Maker Faire weekends and festivals where hundreds and sometimes thousands of rockets are built. Pull out these stands to help your rocketeers with their building and to make sure your PVC building forms don't walk away!

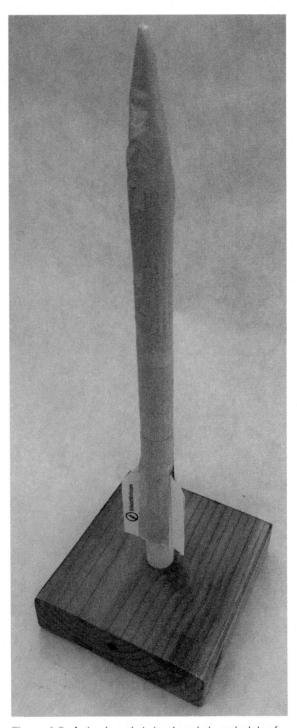

Figure 4-9 *A simple rocket stand can help make lots of rockets.*

Build the Rocket Stand

Step 1

Gather the materials shown in Figure 4-10.

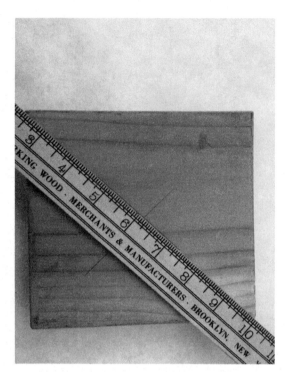

Figure 4-11 *Mark the wood block in the center.*

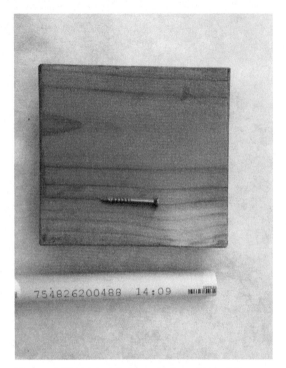

Figure 4-10 *Materials for a single rocket building stand.*

Step 2

Cut a 2"×6"×8' piece of wood into 5-1/2" lengths, yielding about 17 square-ish pieces out of 8' of board. Mark the center of the square, as shown in Figure 4-11, and drill a 7/8" hole on the center to a depth of 1", as shown in Figure 4-12. (Tip: A drill press is really handy for drilling holes of specified depth.)

Figure 4-12 *Use a 7/8" drill bit and drill a 1" deep hole.*

Step 3

Turn the square block on its side and drill a 1/2" hole all the way through so it comes out through the center hole, as shown in Figures 4-13 and 4-14.

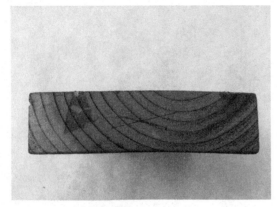

Figure 4-13 *Turn the block on its side and mark the center.*

Figure 4-14 *Use a 1/2" drill bit and drill a hole that intersects the top hole.*

Step 4

Cut a 10' long PVC schedule 40 pipe (1/2" diameter) into 15" lengths. (You'll get eight lengths out of a 10' piece, so cut up two 10'

long pipes if you want to use 16 of the 17 wood squares.)

Step 5

Insert the PVC pipe all the way into the 7/8" hole and then secure it with a 1-5/8" screw inserted through the side hole and screwed all the way through the PVC pipe, as shown in Figure 4-15.

Figure 4-15 *Use a screw to hold the PVC pipe to the block.*

Step 6

With the rocket stand, you now have a tool for creating identical low-pressure rockets over and over again. It's particularly helpful for younger children when it comes time to tape the paper around the PVC to make the rocket body—they don't have to hold on to the PVC and can focus on wrapping the paper around the pipe and taping it securely.

Bounce Rocket

These simple-to-build plastic and rubber rockets go ridiculously high, bounce super high when they hit the ground, and can be built in a few minutes. This project introduces the skill of 3D printing, but fins can be made of other material as well if you don't have access to a 3D printer. You can see an example of a Bounce Rocket in Figure 4-16.

Build the Bounce Rocket

Step 1

Gather the materials shown in Figure 4-17.

Step 2

Cut a 12" length of nylon tube (a nice straight cut) using a hacksaw and use a piece of sandpaper to remove any rough edges around the ends.

Step 3

Print out the Three Fin Guide (*http://bit.ly/ mpgpr_templates*).

Step 4

Mark where the three fins should be placed, as shown in Figure 4-18.

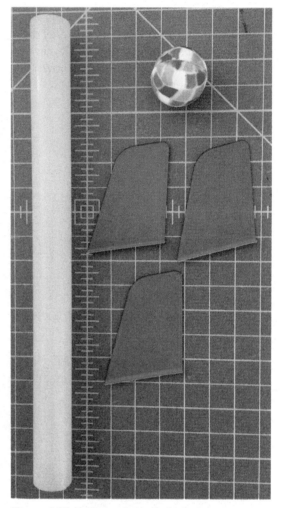

Figure 4-17 *Parts needed to build the bounce rocket.*

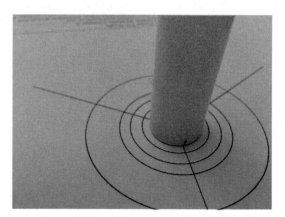

Figure 4-18 *Use the guide to mark where the fins will be placed on the tube.*

Step 5

Wrap a piece of masking tape a distance of 2.5″ from the bottom of the tube where you made the marks for the three fins, as shown in Figure 4-19. Use a piece of fine sandpaper to rough up the portion of the tube below

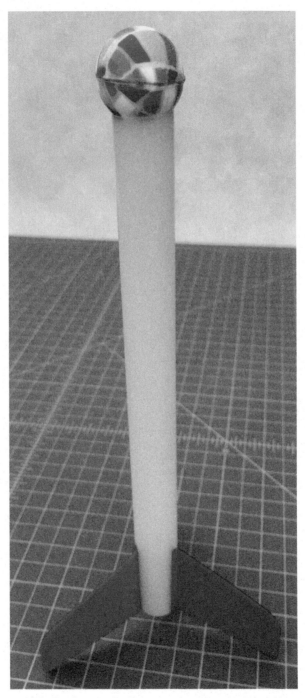

Figure 4-16   *The Bounce Rocket goes up, comes down...and bounces!*

the tape so the fins stick better when you glue them on in a later step.

Step 6

Using a doorjamb to get a straight line, finish the marks made in step 3 around the bottom of the tube. Draw the lines all the way up to the masking tape, as shown in Figure 4-20.

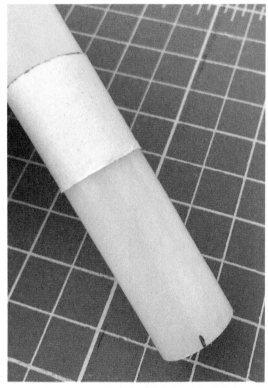

Figure 4-20 *Finish lines from the marks to the masking tape.*

Figure 4-19 *Rough up the area below the tape with sandpaper.*

Step 7

Print three fins on a 3D printer using the STL file (*http://bit.ly/mpgpr_templates*). Figure 4-21 shows the three fins in the 3D printer software being prepared to print.

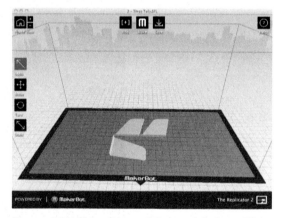

Figure 4-21 *Three fins ready to be printed on a 3D printer.*

Step 8

Using super glue, attach the fins to the side of the body tube along the lines drawn in Step 6. Figures 4-22 and 4-23 show a fin being glued in place.

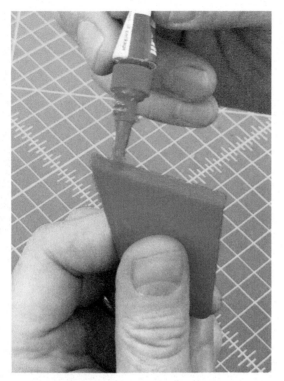

Figure 4-22 *Apply super glue to the short, flattened edge of a fin.*

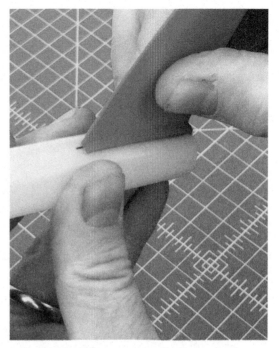

Figure 4-23 *Attach the fins to the lines drawn on the end of the rocket.*

Step 9

Add a line of super glue around the flattened edge of the top of the tube and push the super ball into the glue flow. While holding the ball in place, turn the rocket over and press it down, as shown in Figure 4-24.

Step 10

Use the super glue again and place a bead of glue completely around where the rubber ball and nylon tube meet, as shown in Figure 4-25.

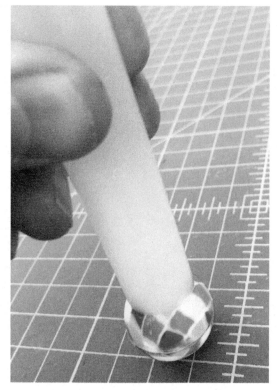

Figure 4-24 *Hold down the super ball until the super glue dries.*

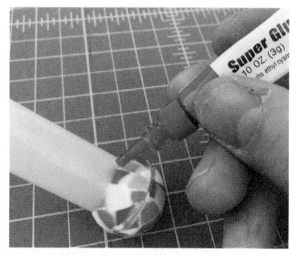

Figure 4-25 *Add some extra super glue to secure the ball to the tube.*

Step 11

Launching! The bounce rocket launches on the 3/8" launch tube that is included in the high-pressure launcher kit (Figure 4-26). This is the same launch tube used for the air rocket glider in Chapter 3; follow the instructions for launching the high-pressure rockets in Chapter 2.

Keep in mind that after a launch this rocket will come down with a little speed. Make certain anyone in the launch area is keeping an eye on this little rocket so the ball doesn't come down on anyone's head. And when it bounces, it could go in any direction, so keep your eyes open!

Figure 4-26 *Bounce rocket on launcher ready to go!*

Air Rocket with Parachute (ARP)

We've been launching air rockets for many years and have been looking for a way to deploy a parachute when the rocket gets to peak altitude. The following is a quick and simple way of going about it using a plastic Easter egg and large craft stick as an "air flap" (Figure 4-27). Our suggestion is to use this as a starting point and then make modifications as you learn new things. With a parachute, you'll need quite a bit bigger launch area than the rockets that just go up and then come right back down.

Build the Air Rocket with Parachute

Step 1

Gather the supplies shown in Figure 4-28. A "Tools and Supplies" list is found at the beginning of the book.

Step 2

Follow the directions in Chapter 2 to build the high-pressure paper and tape rocket. Do everything else, but do not put on the nose cone (Figure 4-29).

Figure 4-27 *This is what the finished ARP rocket looks like.*

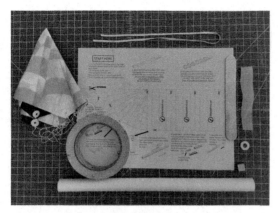

Figure 4-28 *Supplies for ARP.*

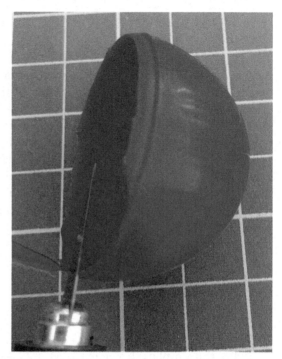

Figure 4-30 *Trim off the edge that holds the plastic egg together.*

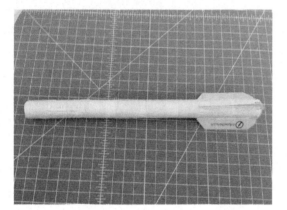

Figure 4-29 *High-pressure paper and tape rocket without nose cone.*

Step 3

Take the bottom (least pointed) half of a medium-size plastic egg and trim off the plastic that holds the egg together (Figure 4-30). Fill the egg half with paper, tape over, and then glue the egg evenly to the rocket body tube with hot glue (Figure 4-31).

Step 4

Next, we'll make the air flap that will hold the contraption closed until deployment at apogee. Trim the top of a large craft stick into a curved shape, then measure down 2 1/2″ (Figure 4-32). Next, shoot a staple through the stick where you made the mark (Figure 4-33). Where the staple came through, bend the staple opposite the curved end of the stick down and then cover in masking tape (Figure 4-34).

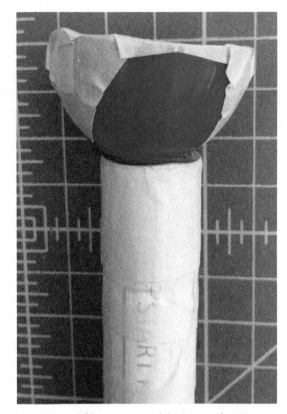

Figure 4-31 Close-up view of the trimmed egg and tape.

Step 5

Cut 8" of rubber band and tie one end to the 1/4" washer (Figure 4-35). This will hook over the staple to keep the parachute deployment contraption closed. Measure down 2" from the egg and tape the craft stick on so it flaps up and down, using the tape as a hinge (Figures 4-36 and 4-37).

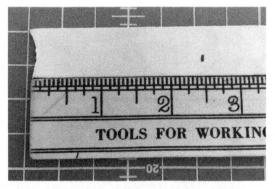

Figure 4-32 Use a measuring stick to make a mark on the craft stick.

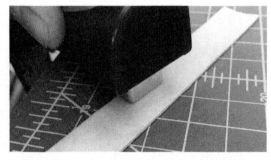

Figure 4-33 Place a staple where you made the mark.

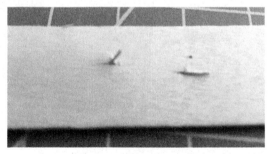

Figure 4-34 Bend one leg of the staple down, and leave the other leg up.

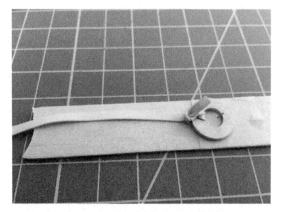

Figure 4-35 *Add a rubber band and washer.*

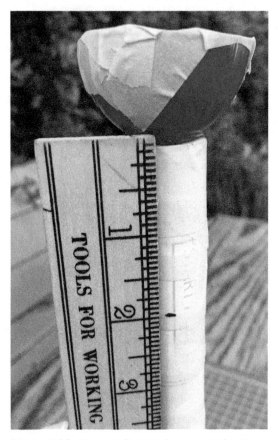

Figure 4-36 *Measure 2" down from egg and make a mark.*

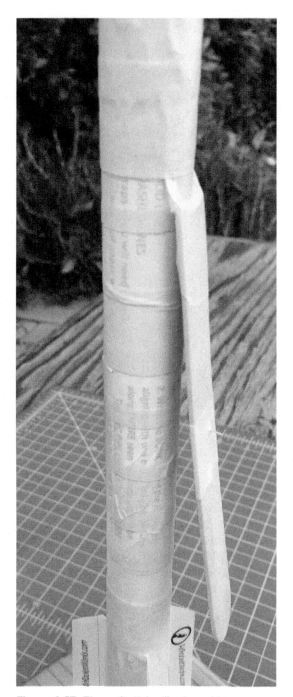

Figure 4-37 *The craft stick will act as a hinge.*

Step 6

Next, we'll make the parachute eject mechanism. Cut a 1" × 4" piece from a plastic milk jug and then fold into an accordion shape, as shown in Figure 4-38. Trace the top half of the egg onto a piece of card stock or manila folder and then cut out on the inside so it is smaller than the diameter of the egg (Figure 4-39). Finally, use tape to hold a small wad of paper in half of the top portion of the egg, then use hot glue to the taped wad and add the card stock disk to the other end (Figure 4-40).

Figure 4-38 *Create the parachute eject mechanism from plastic.*

Figure 4-39 *Trace the egg onto card stock and cut out.*

Figure 4-40 *Attach plastic eject mechanism to egg and card stock.*

Step 7

Now we need to put it all together: place the top half of the egg on top of the bottom. Tape the rubber band (opposite end of the washer) to the top of the rocket body tube opposite of the craft stick wind flap. Stretch a bit and then tape the rubber band to the side of the egg so that the top pops off (Figure 4-41). Then stretch the rubber band over the top of the egg and hook the washer onto the staple on the wind flap (Figure 4-42). When you release the wood wind flap, the egg should pop rapidly open.

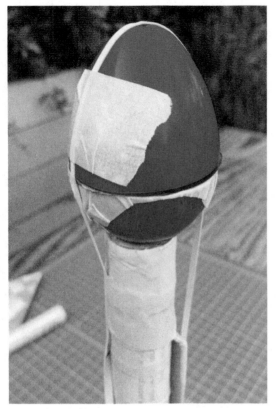

Figure 4-41 *Secure the egg with a rubber band.*

Step 8

At this point, we can add the parachute. Make a simple parachute with a 12" diameter. Tie the ends of the parachute cord together, and then tape to the bottom half of the egg. Fold and pack the parachute, put on the top half of the egg, and stretch the rubber band over it to activate (Figure 4-43). When you release the air flap, the parachute should pop completely out (Figure 4-44).

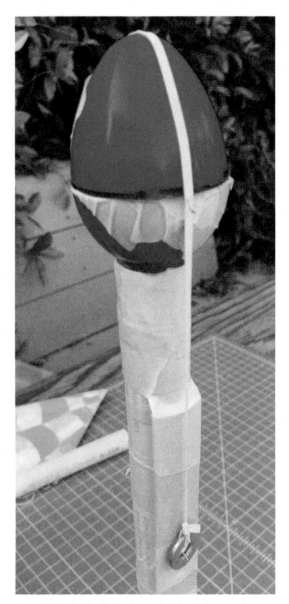

Figure 4-42 *Hook washer onto staple leg.*

Figure 4-43   *Place parachute inside egg.*

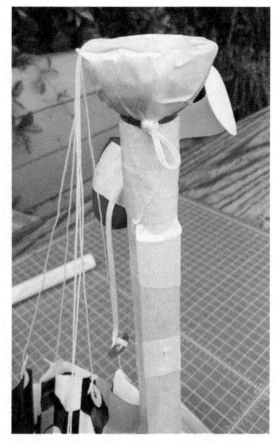

Figure 4-44   *How the parachute will be released.*

Step 9

Cut an 18" piece of coat hanger wire. Wrap it around the 1/2" launch tube and then bend up, as shown in Figure 4-45. This wire will hold the air flap back when the rocket is in the launch position. See the complete rocket in Figure 4-46. As for troubleshooting, if the parachute does not deploy at apogee, tweak it a bit until you get a reliable deployment every time.

Figure 4-45 *The wire will hold down the craft stick hinge.*

Figure 4-46 *Finished rocket with parachute inside egg and ready for launch.*

Step 10

Launch! You'll launch the air rocket with parachute using the high-pressure launcher and the 1/2" launch tube. Because a lot of things are going on at one time with the parachute deployment, you'll have to adjust and modify until you get reliable flights and parachute deployments every time.

Going Further

A maple seed rocket design would fire into the sky and then flutter down like a maple seed. This could be potentially dangerous flying with only one fin, so be careful. See if you can come up with a way to design a maple seed rocket launched by compressed air. Jonathan Mills's Designing a Maple Seed Rocket (*http://bit.ly/maple_seed_rocket*) might get you going on ideas.

Another fun thing to launch into the air is a glider. The air rocket glider in Chapter 3 is a classic example. Other ideas for launching gliders in the air are only limited by your imagination. A number of designs have been employed that use a solid rocket propellant. See if you can come up with something that uses compressed air. "TR-4 Model Rocket Technical Report - Boost Gliders" (*http://bit.ly/boost_gliders*) is a good place to start.

Imagination is the highest kite one can fly.

—Lauren Bacall

With all the focus these days on fancy drones, it's easy to overlook one of the most fun and inexpensive hobbies around: kite flying. With some paper, sticks, glue, and string, you can have something in the air that you can control in less than 20 minutes.

Kites have a long history, with their invention credited to 5th century BC China. In the early days, the frame was created from bamboo, and silk was used for the sail and string. They frequently were made to look like animals (especially flying ones like dragons).

While primarily used as a form of entertainment, kites have also served as signaling devices for early armies, lifting science equipment into the sky, and carrying communication technology such as antennas for radio.

Kites are fun, plain and simple. But their design doesn't have to be plain. (And some kites' construction is anything but simple.) You can find hundreds of kite designs on the Internet, ranging from the basic, four-sided diamond-shaped design to multipiece, laser-cut designs with dozens of parts that move using airflow.

This chapter includes two simple kite projects. The first one can be made with groups in a few minutes. The second, which is much bigger, is made out of Tyvek®, and you can attach a kite camera to it for neat aerial photography.

As you work through the projects, here are some questions that can get a conversation started on kites and how they function:

- What is one factor related to air resistance that all kite builders should remember?

- What is a drawback to a larger kite versus a smaller kite?

- Large kites without strings that can carry one or two passengers exist, but are known by a different name. What are they called?

- Kites come in different shapes. What are the benefits and drawbacks of larger kites (like box kites) over smaller triangular kites?

- What are some interesting uses for kites throughout history?

Quick-Build Kite for $0.25

On a recent trip to the Hawaiian Island of Molokai, Rick met Jonathan, a kite enthusiast who started the Big Wind Kite Factory in 1980. He just goes by Jonathan, and he has been a fan of kites ever since he was young, traveling the hippie trail through Afghanistan. Now he brings his joy of kites to the locals and tourists of Molokai.

Rick and his family bought a little happy face kite and enjoyed browsing Jonathan's workshop, where he makes nearly every kite by hand. Later, Rick's son found some fishing line in the closet of their vacation rental and used it to get that little kite so high it was the size of a pinhead in the brilliant blue sky. While visiting his shop, Jonathan showed Rick a neat, simple little kite design that can be built by 20 kids in 20 minutes (see Figure 5-1). We present Jonathan's 20-minute kite here for individuals or groups of 20, 30 or more!

Build the Quick-Build Kite

Step 1

Gather the supplies shown in Figure 5-2. You can find a complete list of the materials required for this project in "Tools and Supplies".

Figure 5-2 *Quick-build kite supplies.*

Step 2

Fold a 8.5" × 11" piece of paper in half, as shown in Figure 5-3, to create an 8.5" × 5.5" rectangle.

Figure 5-1 *Quick-build kite in flight!*

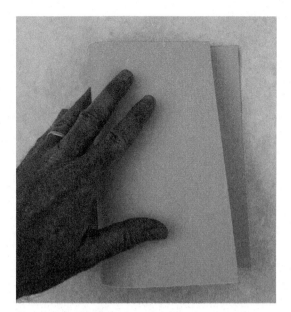

Figure 5-3 *Fold the paper in half.*

Step 3

From the fold on the left side, measure 1″ over on the top and mark; then measure and mark 3″ over on the bottom. Draw a line to connect the two marks, as shown in Figure 5-4.

Step 4

Fold on this newly drawn line, as shown in Figure 5-5.

Step 5

Flip the paper over, as shown in Figure 5-6.

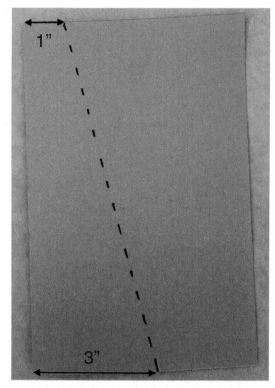

Figure 5-4 *Draw a line between the marks on the top and bottom.*

Step 6

Unfold only the top half of the sheet, and tape along the spine from A to B, as shown in Figure 5-7.

Step 7

Tape a 7.5″ coffee stirrer across the top, as shown in Figure 5-8.

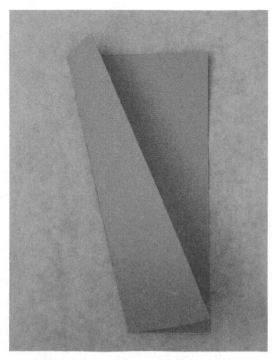

Figure 5-5 *Fold on line.*

Step 8

Underneath the main wings, from the nose of the kite, measure back 2.5″ and place a piece of clear tape (Figure 5-9). This is where the string will go and the tape will provide reinforcement.

Step 9

Punch a hole through the clear tape on the base of the kite and tie on kite string, as shown in Figure 5-10.

Step 10

Wrap the other end of the kite string around a piece of cardboard for your handle. Then cut 3′ of flagging tape—or survey flagging ribbon, as show in the accompanying picture—and tape to the top back of the kite for a tail (Figure 5-11). You're all done and ready to fly this fun little kite!

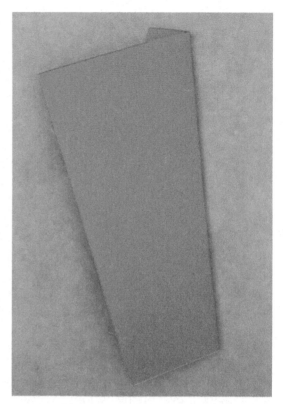

Figure 5-6 *Flip paper over.*

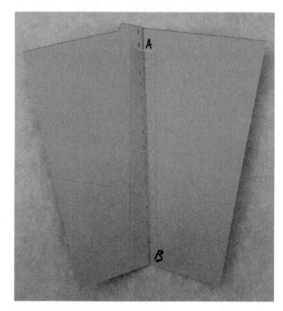

Figure 5-7 *Tape from A to B.*

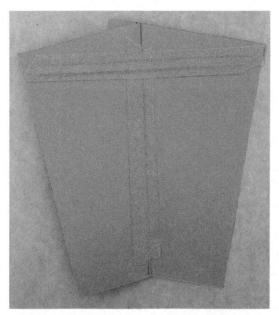

Figure 5-8 *Tape a coffee stirrer across the top of the kite.*

Figure 5-9 *Mark 2.5" from nose of kite.*

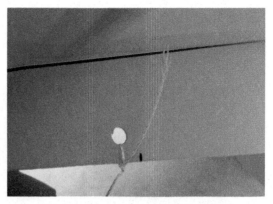

Figure 5-10 *Punch a hold through the tape and tie on your kite string.*

Figure 5-11 *Cut 3' of survey flagging ribbon and tape to the top back of your kite.*

Simple Sled Kite

The quick-build kite is definitely easy to make, and that's a real benefit when you want to get something up in the air fast. But if you've got the time, there's nothing like building a larger, more versatile kite, and we've got a new type of kite for you to explore: the Sled Kite.

Enjoy making this super-basic sled kite, and then flying it almost straight away, in light or moderate winds. No need for tails, unless you would like to add one at the bottom end of each spar (Figure 5-12), just for looks, like we did!

Build the Simple Sled Kite

For this project (Figure 5-12), you can find a complete list of the materials required for this project in "Tools and Supplies".

Step 1

Gather the materials shown in Figure 5-13.

Figure 5-12 *Simple sled kite in the air!*

Figure 5-13 *Materials for the simple sled kite.*

Step 2

Place your Tyvek® sheet flat on a large table. When you do your measuring and cutting, make sure your measurements are accurate and cutting is straight. The kite won't fly well if it's not symmetrical. If you're using 3' wide Tyvek®, measure out and cut at 4' for a 3' × 4' finished piece (Figure 5-14). Then fold the Tyvek® sheet in half so it measures 3' × 2', with the fold on the lefthand side.

Figure 5-14 *The folded Tyvek.*

Step 3

With the fold on the left, starting from the upper-left corner of the folded sheet, measure and mark four dots on the Tyvek®, as indicated in Figure 5-15. Then connect the dots using a ruler and meter stick.

Figure 5-15 *Mark, connect the dots, then cut as shown.*

Step 4

Take your scissors and cut along all the black lines (Figure 5-15). Your sled kite sail is nearly complete!

Step 5

Open up your sail. Now we're going to get ready to add the wooden spars. Lay your 1/4" dowels so they line up with the top sail corner and bottom corners (Figure 5-16).

Step 6

Using Tyvek® tape, tape spars down to the kite sail. Be careful to keep the tape as smooth as possible. You might lay the tape down sticky side up, lay spar on tape (Figure 5-17) and then turn over and tape to kite sail (Figure 5-18). Do this with both spars.

Figure 5-16 *Open up your sail and lay down the 3' long, 1/4" dowels.*

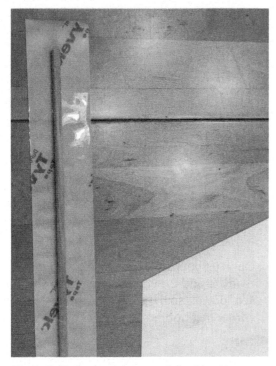

Figure 5-17 *Lay spar down on sticky side of tape then turn over and tape to kite.*

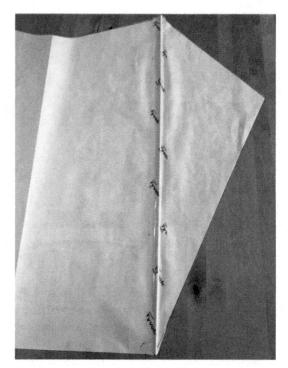

Figure 5-18 *3' spars taped down.*

Step 7

Now we're going work on our two tow points. We're going to be creating Tyvek® tape anchors to attach the bridle (tow line). Starting with the right corner of the sail, stick tape down, as shown in Figures 5-18 and 5-19. Fold the tape over along the edges to create a clean, sharp edge.

Step 8

Now lay out a longer piece of tape on top of the reinforcement you just taped down going out horizontally then fold it over itself, as shown in Figures 5-20 and 5-21.

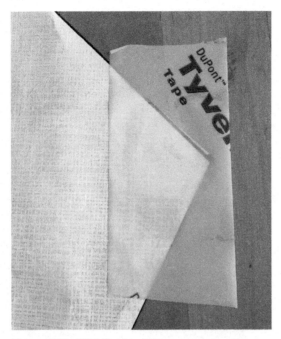

Figure 5-19 *Stick the tape down.*

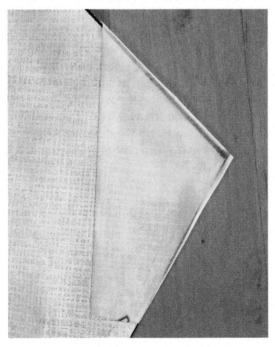

Figure 5-20 *Fold the tape over the edges.*

Figure 5-21 *Longer piece of tape going out horizontally.*

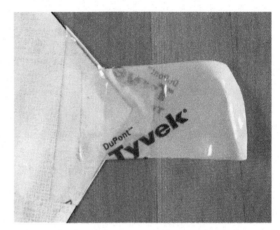

Figure 5-22 *Then fold over onto itself.*

Step 9

Now it's time to attach the bridle. The bridle is the towing point on your kite where your flying line will be tied.

Step 10

The kite is going to have a long bridle, so cut off 15' of line from your roll of kite string. Tie one end of the line tightly to the right side towing point of the kite. (Figure 5-23). Wind the line around twice before tying a knot. Make it as tight as possible to crush the tape.

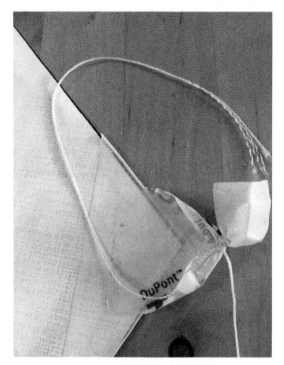

Figure 5-23 *Tie one end of line to right side towing point on kite, and the other end of the 15' bridle line to left side.*

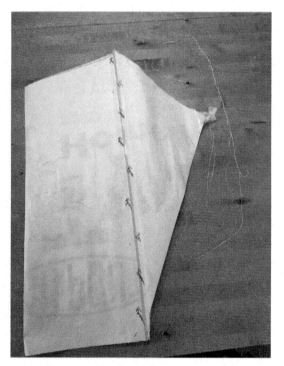

Figure 5-24 *Fold kite and make certain bridle lines are same length.*

Step 11

In the same way, tie the other end of the line to the other towing point on the left.

Step 12

You're almost done! Lay the kite on the table, folded in half so the towing points and spars are directly on top of the other (Figure 5-24). Stretch out the bridle lines and tie a standard knot right near the end where the lines come together. The two bridle lines should be exactly the same length. The photo in Figure 5-24 shows the end brought around so you can see how it should be tied. Your flight line will be then tied securely to the bridle loop.

Step 13

Flying! With the help of a friend holding the sled kite and facing the wind, let the wind fill the kite and then let it go. If you want, add two lengths of nonadhesive flagging tape to the bottom points of the kite just for fun.

Step 14

Try using a cordless drill to wind your kite back in!

Going Further

My Best Kite (http://www.my-best-kite.com/)

Australian Tim Parish is the founder of My Best Kite (*http://www.my-best-kite.com*). He's got lots of free kite plans on his site, from simple to sophisticated. From delta and box kites to diamond and dragon, you can find loads of information there to keep you going with kite flying for some time.

Build a Kite Aerial Photography Rig (http://bit.ly/kite_photo_rig)

Add a mini camera to your kite for a cool eye in the sky! This is a great one to build, and it also includes plans for a delta kite.

About the Authors

Rick Schertle has taught middle school language arts in San José, California, the past 22 years and writes occasionally for *Make: Magazine*. His diverse interests include world travel, backpacking, tinkering, and all things that fly—all more fun with the enthusiastic support of his wife and crazy antics of his son and daughter. His original project was the Air Rocket Launcher for *Make:* issue 15 and since then has brought air rocket fun to Maker Faires and events all over the world.

James Floyd Kelly is a full-time writer in Atlanta, Georgia. James has written over 30 books on a mix of technology subjects that include building your own CNC machine and 3D printer, game programming, LEGO robot design, Computer-Aided Design (CAD), and open source software. He and his wife have two young boys who are showing solid signs of possessing the maker gene.

Colophon

The cover, body, and header font is Benton Sans.

CPSIA information can be obtained
at www.ICGtesting.com
Printed in the USA
BVOW10n0327190516

448727BV00001B/2/P

9 781457 187698

Make: Planes, Gliders, and Paper Rockets

Do helicopters need more or less energy to stay in the sky than an airplane? What pushes a rocket to leave the atmosphere? Why can airplanes have smaller motors than helicopters?

Help your students learn the answers to these and other questions!

Written for educators, homeschoolers, parents—and kids!—this fully illustrated book provides a fun mix of projects, discussion materials, instructions, and subjects for deeper investigation around the basics of homemade flying objects. With the projects in this book, you can spend more time learning and experimenting, and less time planning and preparing.

Complete with download links to PDF templates that expand your teaching, this is your one-stop manual for learning about, interacting with, and being curious about airflow, gravity, torque, power, ballistics, pressure, and force.

In *Make: Planes, Gliders, and Paper Rockets*, you'll make and experiment with:

» **Paper catapult helicopter—add an LED light for night launches!**

» **Pull-string stick helicopter**

» **Rubber band airplane**

» **Simple sled kite**

» **25-cent quick-build kite**

» **Air rockets with a parachute or a glider**

» **Foam air rocket**

» **Rocket stands**

» **Bounce rocket**

» **Low- and high-pressure rocket launchers**

What will you put in the air?

US **$19.99** CAN **$22.99**
ISBN: 978-1-4571-8769-8

Make:
makezine.com